DIGESTIVE WELLNESS

Elizabeth Lipski

M.S., C.C.N.

KEATS PUBLISHING, INC.

New Canaan, Connecticut

Digestive Wellness is not intended as medical advice. Its intent is solely informational and educational. Please consult a health professional should the need for one be indicated.

Library of Congress Cataloging-in-Publication Data

Lipski, Elizabeth.
Digestive wellness / by Elizabeth Lipski.
p. cm.
Includes bibliographical references.
ISBN 0-87983-679-2
1. Digestion—Popular works. 2. Indigestion—Popular works.
OP145.L57 1996
616.3—dc20
95-49689
CIP

Printed in United States of America

Published by Keats Publishing, Inc.
27 Pine Street, Box 876
New Canaan, Connecticut 06840-0876

98 97 96 6 5 4 3 2 1

Digestive Wellness

This book is dedicated to my mother and father
who taught me that I could fly
and to my husband
who's been soaring beside me.

Acknowledgments

My gratitude and thanks to:

My husband, Stephen Long, who opened the space in my life so I could write this book and for his many devoted hours of editing my work.
My children, Kyle and Arthur, who missed me.
Gabrielle Resnick who kept my life in order.
Stephen Barrie, N.D., and Marty Lee, Ph.D., who generously gave me access to information and illustrations for my book.
Russell Jaffe, M.D., who helped me shape my ideas in the earliest phase and whose ideas have greatly influenced my work.
Jeff Bland, Ph.D., who was my first nutrition teacher and continues to amaze me with the depth of his knowledge decades later.
Lori Feiss for drawing the illustrations.
Nathan Keats and Don Bensen at Keats Publishing for putting their trust in an unknown author.
Susan E. Davis for shaping my manuscript into a real live book.
And to the many people who shared their knowledge, time, and work with me: Leo Galland, M.D., Corey Resnick, N.D., Bill Shaddle, Bob Smith, Natasha Trenev and Billy Crook, M.D.

Disclaimer

The therapies in this book have been helpful for many people and have been scientifically researched and documented. Because I do not know you or your particular needs, they may or may not be appropriate for you. This book is not intended to displace proper medical care. Do not stop taking medications without discussion with your physician. If you want to try the approaches discussed, take this book to your physician and ask him/her to help you by ordering the appropriate laboratory tests and monitoring your progress. Teamwork can often bring the best results.

CONTENTS

PART 5
A FUNCTIONAL APPROACH TO THERAPEUTICS/SELF CARE

PART 6
RESOURCES

Part I

Digestive Health

Chapter 1

WHO CAN BENEFIT FROM THIS BOOK?

"If the patient has been to more than four physicians, nutrition is probably the medical answer."

ABRAHAM HOFFER, M.D., PH.D.

THERE IS CURRENTLY an epidemic of digestive illness in our country, an epidemic that is directly related to the foods we eat and the way we live. One-third to one-half of all adults have digestive illness—over 62 million people. According to a May 1994 study, 69 percent of the people studied reported having at least one gastrointestinal problem within the previous three months. Except for the common cold, digestive problems are the most common reason people seek medical advice. They are the third largest category of illness in the United States, at a cost of $41 billion, and result in 229 million days of lost productivity, at a cost of $14.5 billion per year.[1] Constipation plagues our nation. Americans each average over $500 a year on laxatives alone. Year after year, medications for digestive illness top the pharmaceutical best-seller list. Zantac, an ulcer medication, is the best selling drug of all time, with sales of well over $1 billion annually.

When I first began studying nutrition, I remember going to meetings where the old-time nutritionists would lecture in evangelistic style, foretelling "death begins in the colon" while pointing to posters of enormous colons. At the time I thought they were being overly dramatic, but I have slowly come to learn that they were right. For many of us, disease and death do begin in the digestive system.

Most of us don't think much about digestion unless it isn't working well. We don't have to because it works automatically. The function of digestion is to break foods down into basic components for the cells to use for energy, as building materials and as catalysts. The uninterrupted flow of these nutrients into our system is critical to our long-term health. When we eat poorly or our digestion becomes blocked and sluggish, we compromise the ability of *all* our cells to work efficiently and healthfully. While it may seem obvious to some that what we eat affects the health of our digestive systems and our bodies, it is a revolutionary idea to many. There is a consensus of agreement among health organizations and professionals that we can reduce the risk of heart disease, diabetes and cancer by eating more healthful foods. But many people with chronic digestive problems continue to eat poorly, never realizing that their food choices are causing their ill health.

This book is a resource for people with faulty digestion and the people who care about them. Some of you may have serious digestive illness, like Crohn's disease, irritable bowel syndrome or ulcerative colitis, while others may have simpler problems of heartburn, indigestion or constipation. In addition to obvious digestive problems, many other medical problems are caused by faulty digestion. Arthritis, many auto-immune diseases, eczema, food sensitivities and psoriasis are digestive in origin. People with migraine headaches nearly always have food sensitivities that trigger the problem. Chronic fatigue syndrome, while initially triggered by a virus, has a major digestive component as well. When digestive competence is restored, many of these problems are resolved.

ANNUAL BURDEN OF SELECTED DIGESTIVE DISEASES IN THE UNITED STATES

Disease	**Number of people with this condition, in millions**	**Days of restricted activity, in millions**	**Deaths caused by this condition**
GI infections	8.3	22.8	456
GI cancers	0.258	14	79,854
Hemorrhoids	10.36	8.2	22
Esophageal disease	0.546	2.4	1,600
Ulcer	4.491	28.8	6,715
Gastritis & duodenitis	2.793	7.2	641
Hernia	4.741	32.5	1,391
Inflammatory bowel disease	2.308	20.3	628
Diverticular disease	1.9	9.2	2,933
Constipation	4.458	1.9	31
Irritable bowel	1.379	2.2	63
Chronic liver disease	0.552	10.6	32,107
Gallstones	0.956	8.8	2,975
Other GI conditions	18.958	60.1	61,584
Total GI conditions	**62**	**229**	**191,000**

Digestive Diseases in the United States: Epidemiology and Impact, U.S. Department of Health and Human Services, Public Health Service, National Institutes of Health, Publication No. 94-1447, May 1994, p.19.

DIGESTIVE HEALTH PROBLEMS		
Mouth ulcers	Tongue problems	Periodontal disease
Heartburn	Belching	Hiatal hernia
Ulcers	Gastritis	Bloating & gas
Irritable bowel syndrome	Crohn's disease	Ulcerative colitis
Celiac disease/ Sprue	Gallbladder problems	Indigestion
Constipation	Diarrhea	Diarrhea & constipation
Chronic fatigue	Food sensitivities	Arthritis, all types
Yeast infections	Migraine headaches	Ankylosing spondilitis
Eczema	Psoriasis	

New medical research and clinical studies indicate that digestive illness is often caused by things we can control:

Poor food choices
Lack of dietary fiber
High levels of daily stress
Pharmaceuticals, both over-the-counter and recreational drugs
Alcoholic beverages
Cigarette smoking
Environmental irritants and toxins

In fact, the only factor we can't control is our genetic make-up.

Dennis Burkitt, father of the fiber revolution, found almost no appendicitis, colon disease, diabetes or hiatal hernia in people eating traditional diets. When these people move to cities or change to a Westernized diet—of high sugar, highly processed, low-fiber and low-nutrient-density foods—they develop these illnesses in the same amounts as Westerners do. Other studies show that ulcerative colitis and Crohn's disease are increasing in cultures that switch to a Western diet.[2]

In the last five years, research has yielded surprising results, turning the arena of digestive health on its head. Ulcers, long believed to be a disease of high stress and high stomach acidity, are now known to be caused by bacteria. Of people with ulcers, 79 to 90 percent have a bacteria called *Helicobacter pylori*. When it takes hold, the mucus layer of the stomach is digested, and the acid comes into direct contact with the unprotected stomach lining, causing it to burn and develop sores called ulcers. Though Zantac and Tagamet, the most commonly used medications for ulcers, are the best selling drugs of all time, the latest recommendations from the National Institutes of Health put these medications in the background and focus instead on killing the *Helicobacter* bacteria.[3]

Ulcers haven't been the only surprises in the field of digestive health. Unfortunately, we often lose some of our ability to function long before a disease is present. Now medical laboratories are developing new tests that focus on function rather than disease. They are able to give us information that can serve as an early warning device, letting us know when our organs are losing their abilities. These labs are asking new questions and getting new answers that are changing internal medicine.

Innovative stool testing has shown that about a quarter of the people tested harbor parasites. Some may be harmless, while others cause illness that resolves when the parasitic infestation is treated. Parasites cause a wide range of symptoms, including gastrointestinal complaints of constipation, diarrhea, gas and bloating, as well as allergies, anemia, fatigue, immune system problems, joint and muscle aches and pains, nervousness, skin conditions and sleep disturbances.

Research on microbes provides a new view of the symbiosis of our own bodies and the organisms dwelling there. Now we can look inside without invasive techniques to see which microbes are present and how well we are digesting food. There are 10 times more bacteria in our intestinal tract than cells in our body. They do important work like producing vitamins, digesting the sugar in milk and keeping our intestines acidic, which protects us from disease-producing viruses and bacteria. Antibiotics, stressful living conditions and poor dietary habits destroy this balance, and opportunistic microorganisms can take over. We call this imbalance "intestinal dysbiosis."

These microorganisms can also travel out of the digestive tract and create havoc throughout the body. One theory especially intrigues me: displaced bacteria, fungi or parasites can set up a local immune reaction that mimics specific gene markers. The immune system sees the gene markers as something to attack and autoimmune diseases develop. It is believed that this mechanism underlies some serious diseases like rheumatoid arthritis and ankylosing spondylitis.

We are becoming more sensitive all the time to the foods we eat. Leaky gut syndrome (also called intestinal permeability), another new concept in digestive health, plays a large role in these sensitivities. Our intestines have a dual function: to let nutrients into our bloodstream while blocking the absorption of other materials. When foods aren't digested properly, they ferment and rot in our intestines, producing toxic byproducts that irritate and inflame the intestinal lining, causing tiny tears. The intestinal lining loses its ability to act properly as a filter and leaks, hence the name leaky gut syndrome. It is a contributing factor in Crohn's disease, celiac disease, irritable bowel and ulcerative colitis. It is often responsible for illness in distant tissues and organs, including problems like chronic fatigue syndrome, food sensitivities, eczema, migraine headaches, psoriasis and osteoarthritis.

Digestive Wellness provides you with a step-by-step plan for making healthful changes in your lifestyle. The approach is from a biological rather than a medical viewpoint. The standard medical approach is to diagnose and provide "appropriate" treatment, either drugs or surgery. The biological approach involves cleansing, feeding, and nurturing your entire being—simple but effective tools to improve the way you feel. By understanding the function of the various parts of the GI (gastrointestinal) tract and looking for underlying causes of disease rather than mere treatment of symptoms, we can begin to understand how to correct our problems.

In the first part of the book we explore the causes of digestive illness due to the American lifestyle. Then we take a trip through the digestive tract where we find a beautifully orchestrated system of integrated harmony. Then we look at the microbes that populate our digestive universe. When they are out of balance, we feel the effects. We move on to discuss dysbiosis and leaky gut syndrome, which often underlie digestive illness and many seemingly unrelated health problems. These chapters provide the groundwork so you can really understand the causes and effects of poor lifestyle choices and medical therapies on your condition.

In the next part we move on to self-improvement, with information and practical tips on how to develop a wellness lifestyle. Information about exercise, stress reduction programs, food choices and shopping lists are provided. We'll focus on a personal exploration of what makes you feel better or worse and recommend approaches that support your body's ability to evoke its natural healing response. You will discover which habits make you feel vibrant and energetic and which drain you. The biological approach, which treats you as a whole person with unique needs, is based on the concept of "biochemical individuality." Just as each of us has a unique face, body and personality, so too do we each have a unique biochemistry. One person's need for a specific nutrient, for example, can be 30-fold higher or lower than another person's. When it comes to food, one person's pleasure is another's poison. Although you may believe that it's important to eat certain foods—like bread, eggs, meat, milk, oranges and tomatoes—they may or may not be healthy for you. It depends on how well *your* body can use them. Detoxification programs are discussed: why you need to incorporate them in your life, recommended programs and what to expect during detoxification.

Functional medicine comes next, which is concerned with early intervention in health problems. Early intervention improves your chances of returning to full health. Information on functional lab tests is detailed. This information is new, and most physicians will be unfamiliar with many of these laboratory tests. Take this book to your doctor's office, and ask your doctor to work with you in this new way.

The next part of the book discusses self-care strategies for digestive and digestively related health problems, including information about the latest research on nutritional and herbal therapies. This part is divided into two chapters. The first discusses common digestive illnesses and conditions, and the second explores problems that are the consequence of faulty digestion, like arthritic conditions, migraine headaches and skin problems. Research has been gathered from clinicians and researchers who are striving to learn how your body works and why it fails. The goal is to help your body reach its own natural balance, which will allow it to heal. Day in and day out, cell by cell, your body continually replaces itself. With the correct balance of work, rest and nutrients, your body can become healthier each year. If you build your "house" with excellent materials, it will wear better!

Finally, a resource guide is included. It lists professional organizations that can refer you to nutritionally oriented physicians, health professionals and laboratories. For those who want to read more on a specific topic, a suggested reading list is also provided.

Throughout the book are exercises and questionnaires designed to increase your self-awareness of mind and body, help you shop more wisely, breathe more deeply, relax more fully and live more freely. Even though we may not be aware of it, we all practice medical self-care. When we get a headache, we take an aspirin or go for a walk. If we have indigestion, we take an Alka-seltzer or drink ginger tea. We know when we're too sick to go to work. Most of the time, we make our own assessment and treatment plan, expecting that the problem will pass with time. When these plans fail, we seek professional help. This book will expose you to more plans, new ideas and the tools to be your own health expert. Just as one tool won't work for every job, not all of these tools will work for you. But some will, and even the failures may give you useful information.

The good news is that you can change the way you feel. The bad news is that it takes work and personal commitment. The good news about the bad news is that the journey can be an amazing voyage of self-discovery and self-mastery. This book provides you with some of the tools you need. In fact, the voyage is as interesting as the destination. So work, relax, laugh and remember to look out the window and enjoy the scenery. This book is about taking control of your lifestyle to increase your chances of getting healthier and more vibrant each year.

Exercise 1

DIGESTIVE HEALTH APPRAISAL QUESTIONNAIRE[4]

DATE: ____________________________

This questionnaire will help you assess your digestive status. It is not meant as a replacement for a physician's care. The answers will help you focus your attention on specific areas of need.

MEDICATIONS USING CURRENTLY

Circle any of the following medications you are taking. Write down the dosage and frequency:

Antacids	Laxatives	Cortisone
Antibiotics	Oral contraceptives	Prednisone
Antifungals	Ulcer medications	Tylenol
Anti-inflammatories	Aspirin	Stool softeners

OTHER ____________________________

FOOD, NUTRITION AND LIFESTYLE

Circle if you eat, drink or use:

Alcohol	Luncheon meats
Candy	Margarine
Cigarettes	Soft drinks
Coffee	Sweets/pastries
Fast foods	Chew tobacco
Fried foods	

Circle if you:

Diet often
Do not exercise regularly
Are under excessive stress
Are exposed to chemicals at work
Are exposed to cigarette smoke

This part of the questionnaire will help you discover where your digestive system is having problems. It is a screening tool and does not constitute an exact diagnosis of your problem. However, it can point you in the right direction in determining where the highest priorities lie in your healing process.

Instructions: Circle the number which best describes the intensity of your symptoms. If you do not know the answer to a question, leave it blank. Add the totals for each section to assess which areas need your attention.

0 = Symptom is not present/ or rarely
1 = Mild/sometimes
2 = Moderate/often
3 = Severe/almost always

Part 1

SECTION A:
HYPOACIDITY OF THE STOMACH

1	Burping	0	1	2	3
2	Fullness for extended time after meals	0	1	2	3
3	Bloating	0	1	2	3
4	Poor appetite	0	1	2	3
5	Stomach upsets easily	0	1	2	3
6	History of constipation	0	1	2	3
7	Known food allergies	0	1	2	3
	Total:				

Score 0-4: Low priority **Score 5-8: Moderate priority**
Score 9+: High priority

SECTION B:
HYPOFUNCTION OF SMALL INTESTINES AND/OR PANCREAS

1	Abdominal cramps	0	1	2	3
2	Indigestion 1 to 3 hours after eating	0	1	2	3
3	Fatigue after eating	0	1	2	3
4	Lower bowel gas	0	1	2	3
5	Alternating constipation & diarrhea	0	1	2	3
6	Diarrhea	0	1	2	3
7	Roughage & fiber causes constipation	0	1	2	3
8	Mucus in stools	0	1	2	3
9	Stool poorly formed	0	1	2	3

HYPOFUNCTION OF SMALL INTESTINES AND/OR PANCREAS (con't)

10	Shiny stool	0	1	2	3
11	Three or more large bowel movements daily	0	1	2	3
12	Dry, flaky skin &/or dry brittle hair	0	1	2	3
13	Pain in left side under rib cage	0	1	2	3
14	Acne	0	1	2	3
15	Food allergies	0	1	2	3
16	Difficulty gaining weight	0	1	2	3
17	Foul-smelling stool	0	1	2	3
	Total:				

Score 0-4: Low priority **Score 5-8: Moderate priority**
Score 9+: High priority

SECTION C:
ULCERS/HYPERACIDITY OF THE STOMACH

1	Stomach pains	0	1	2	3
2	Stomach pains just before or after meals	0	1	2	3
3	Dependency on antacids	0	1	2	3
4	Chronic abdominal pain	0	1	2	3
5	Butterfly sensations in stomach	0	1	2	3
6	Difficulty belching	0	1	2	3
7	Stomach pain when emotionally upset	0	1	2	3
8	Sudden, acute indigestion	0	1	2	3
9	Relief of symptoms by carbonated drinks	0	1	2	3

ULCERS/HYPERACIDITY OF THE STOMACH (con't)

10	Relief of stomach pain by drinking cream/milk	0	1	2	3
11	History of ulcer or gastritis	0	1	2	3
12	Current ulcer	0	1	2	3
13	Black stool when not taking iron supplements	0	1	2	3
	Total:				

Score 0-4: Low priority **Score 5-8: Moderate Priority**
Score 9+: High priority

SECTION D: COLON/LARGE INTESTINE

1	Seasonal diarrhea	0	1	2	3
2	Frequent and recurrent infections (colds)	0	1	2	3
3	Bladder and kidney infections	0	1	2	3
4	Vaginal yeast infection	0	1	2	3
5	Abdominal cramps	0	1	2	3
6	Toe and fingernail fungus	0	1	2	3
7	Alternating diarrhea/constipation	0	1	2	3
8	Constipation	0	1	2	3
9	History of antibiotic use	0	1	2	3
10	Meat eater	0	1	2	3
11	Rapidly failing vision	0	1	2	3
	Total:				

Score 0-4: Low priority **Score 5-8: Moderate priority**
Score 9+: High priority

SECTION E: LIVER/GALLBLADDER

1	Intolerance to greasy foods	0	1	2	3
2	Headaches after eating	0	1	2	3
3	Light-colored stool	0	1	2	3
4	Foul-smelling stool	0	1	2	3
5	Less than one bowel movement daily	0	1	2	3
6	Constipation	0	1	2	3
7	Hard stool	0	1	2	3
8	Sour taste in mouth	0	1	2	3
9	Gray-colored skin	0	1	2	3
10	Yellow in whites of eyes	0	1	2	3
11	Bad breath	0	1	2	3
12	Body odor	0	1	2	3
13	Fatigue and sleepiness after eating	0	1	2	3
14	Pain in right side under rib cage	0	1	2	3
15	Painful to pass stool	0	1	2	3
16	Retain water	0	1	2	3
17	Big toe painful	0	1	2	3
18	Pain radiates along outside of leg	0	1	2	3
19	Dry skin/hair	0	1	2	3
20	Red blood in stool	No			Yes
21	Have had jaundice or hepatitis	No			Yes
22	High blood cholesterol and low HDL cholesterol	No	Unknown		Yes
23	Is your cholesterol level above 200?	No	Unknown		Yes
24	Is your triglyceride level above 115?	No	Unknown		Yes
	Total:				

Score 0-2: Low priority **Score 3-5: Moderate priority**
Score 6+: High priority

SECTION F:
INTESTINAL PERMEABILITY/LEAKY GUT SYNDROME, DYSBIOSIS

1	Constipation and/or diarrhea	0	1	2	3
2	Abdominal pain or bloating	0	1	2	3
3	Mucus or blood in stool	0	1	2	3
4	Joint pain or swelling, or arthritis	0	1	2	3
5	Chronic or frequent fatigue or tiredness	0	1	2	3
6	Food allergy or food sensitivities or intolerance	0	1	2	3
7	Sinus or nasal congestion	0	1	2	3
8	Chronic or frequent inflammations	0	1	2	3
9	Eczema, skin rashes or hives (urticaria)	0	1	2	3
10	Asthma, hayfever or airborne allergies	0	1	2	3
11	Confusion, poor memory or mood swings	0	1	2	3
12	Use of nonsteroidal anti-inflammatory drugs (aspirin, Tylenol, Motrin)	0	1	2	3
13	History of antibiotic use	0	1	2	3
14	Alcohol consumption, or alcohol makes you feel sick	0	1	2	3
15	Ulcerative Colitis, Crohn's disease or celiac disease	0	1	2	3
	Total:				

Score 1-5: Low priority **Score 6-10: Mild case**
Score 7-19: Moderate priority **Score 20+: High priority**

SECTION G:
GASTRIC REFLUX

1	Sour taste in mouth	0	1	2	3
2	Regurgitate undigested food into mouth	0	1	2	3
3	Frequent nocturnal coughing	0	1	2	3
4	Burning sensation from citrus on way to stomach	0	1	2	3
5	Heartburn	0	1	2	3
6	Burping	0	1	2	3
7	Difficulty swallowing solids or liquids	0	1	2	3
	Total:				

Score 0-3: Low priority **Score 4-6: Moderate priority**
Score 7+: High priority

INTERPRETATION OF QUESTIONNAIRE

MEDICATIONS

- Medications are good indicators that your body is in some sort of imbalance.
- Medications have drug/nutrient interactions. Some nutrient needs may be increased, some decreased, some nutrients may block absorption or usefulness of the drug.

FOODS, DRINKS, TOBACCO

- Candy, alcohol, sweets and soft drinks: These "empty calorie foods" contain few nutrients, but nutrients are needed to metabolize them, and they replace healthy foods in our diets. These foods have a detrimental effect on most digestive problems; for instance, simple sugars feed Candida, bacteria and parasites.

- Cigarettes and chewing tobacco: Make sure to take a good antioxidant supplement and lots of vitamin C to compensate for the stress the tobacco causes. Tobacco has a negative effect on the digestive system.
- Luncheon meats, pastries, fast-foods and margarine: If you eat these foods you are probably getting too much fat, especially saturated fat. Margarine and most pastries also contain "hydrogenated oils," which are absorbed into our cells but are detrimental to our health. They make the cell membranes stiff and stifle the intake of nutrients and outgo of wastes, promote free radical activity, and contribute to atherosclerosis and inflammatory diseases.

Lifestyle

- Diet often: Weight problems can be caused by a hypoactive thyroid, food sensitivities, poor food choices, sedentary lifestyle and emotional and social overeating. Chronic dieting leads to further metabolic slowdown. A wellness-centered approach works best for the overweight person.
- Lack of routine exercise: Exercise is the great stress reducer and enhances the health of our whole body, including our digestive system. Regular exercise at least three times a week for 20 to 30 minutes can significantly reduce the risk of cardiovascular disease and increase our total sense of well-being.
- High stress level: This indicates the need for a good exercise program, ways of nurturing oneself and training to increase emotional heartiness. Food choices usually

suffer during stressful periods, while nutrient needs are increased. Supplementation may be indicated.

- Exposure to chemicals: Prolonged exposure to chemicals can cause environmental illness, which can manifest as obvious illness or as nondiagnosable complaints of confusion, chronic fatigue, headaches or just not feeling right. Many women with breast cancer have had prolonged exposure to chemicals. Metabolic clearing and low-temperature saunas are important.
- Exposure to cigarette smoke: Research indicates that second-hand smoke is detrimental to a healthy respiratory system. If you cannot get away from smokers, buy them "smokeless" ashtrays, open windows whenever possible and take antioxidant supplements.
- Focus your attention on the sections where you scored in either the moderate or high priority range. They are the greatest arenas for health enhancement of your digestive system.

Chapter 2

THE AMERICAN WAY OF LIFE IS HAZARDOUS TO YOUR HEALTH

"Of the 10 leading causes of death in the United States, four, including the top three, are associated with dietary excess: coronary heart disease, some types of cancer, stroke, non-insulin-dependent diabetes mellitus. Together these conditions account for nearly two thirds of the deaths occurring each year in the United States."

BETTY FRZAZO, "THE HIGH COST OF POOR DIETS," *USDA FOOD REVIEW*, JAN.-APRIL 1994,17(1): 2

WE EAT VERY differently than our great-grandparents did. Food production began to change during the Industrial Revolution, making refined sugar and white flour, previously a luxury of the rich, affordable and available to everyone. However, in processing, fiber and nutrients are lost. This was just the beginning of the changes introduced into our food supply—frozen foods, packaged foods, microwave ovens, foods shipped globally! Today we are part of a massive, uncontrolled science experiment. What happens when people are fed highly processed foods, lacking in nutrients and fiber and loaded with chemicals, over three generations? What happens when you put these same people under high levels of stress, in sedentary jobs, with poor air and water quality? Is it a coincidence that infertility rates are up, that Americans are fatter than ever before, that we are more violent than ever before, and that more people are committing suicide? Is it a coincidence that more children have attention deficit syndrome, a condition that never existed previously, and that children and adults have more allergies and chronic ear infections? Is it a coincidence that our immune systems are breaking down or that diabetes and heart disease rates have changed dramatically over the past 80 years? I don't think so.

According to the United States Department of Agriculture (USDA) in 1988-1989, Americans each ate on average:

365 servings of soda pop or 638 cans per year for people aged 12-29
134 pounds of refined sugar excluding honey
90 pounds of fats and oils
63 dozen donuts
60 pounds of cakes and cookies
23 gallons of ice cream
22 pounds of candy
8 pounds of corn chips, popcorn and pretzels
7 pounds of potato chips

If you add up the calories we consume each day of high-calorie, nutrient-poor foods, *nearly half of our caloric intake comes from nutritionally depleted foods.* In addition, we also drink 2.65 gallons of pure alcohol per person each year, which equals 50 gallons of beer, 20 gallons of wine, or more than 4 gallons of distilled liquor.[1] No wonder, the *S*tandard *A*merican *D*iet is SAD!

For the first time in 1978, a U.S. Senate subcommittee chaired by George McGovern made recommendations about healthy eating. They proposed cutting fat intake by a third, doubling fiber intake, eating at least 2½ cups of produce daily, focusing on whole grains and legumes, viewing meat as a sidedish rather than the main course, integrating more vegetarian meals, and cutting sodium. Lobbyists from the food industry were up in arms and defended the status quo.

In 1993, the USDA released a new food pyramid, which is an updated but watered down "improvement" of the 1978 recommendations. The numbers have been removed, so they won't offend anyone, but that made the guidelines ambiguous. Bland as they were, even this provoked controversy. The meat and dairy lobbies exerted their clout so that meat and dairy play a more prominent part in a daily menu than is optimal. The resulting pyramid provides a new shape, but offers little change to the old four basic food groups.

Even disguised, the U.S. Dietary Guidelines were heard loud and clear by the media and food manufacturers. You'd have to live in a cave to be unaware of why low-fat, fat-free foods are on the market. Clever packaging claims—"no cholesterol" or "low fat"—sway purchasing decisions. But despite increased awareness of what we ought to eat, Americans still don't eat right. According to the latest USDA food consumption study, though many of us seem to understand the concepts, we are still not making healthy choices. For example, we eat 10 percent fewer vegetables than we did just 10 years ago, and in the same decade use of soft drinks has grown by 60 percent. We're consuming more sugar—a whopping 135 pounds per person per year, up from 126 pounds in 1976 and only 15 pounds in 1850. Only 9 percent of the population eats five servings of fruits and vegetables a day.[2]

Though Americans have significantly decreased the amount of fat in their diets, only a quarter of us meet the recommendations to reduce fat to 30 percent of total calories. Even at that, many researchers believe the ideal level is less than 20 percent. High-fat diets are a primary cause of gallbladder and heart diseases, and they contribute to obesity, which raises the risk of all degenerative illnesses. Of women aged 30 to 74, 30 to 35 percent are overweight and getting fatter. Almost as many men are overweight. We weigh more now than we did just 10 years ago.

Even though most Americans still eat too much fat, we also eat the wrong kind. In 1910 a process called hydrogenation was invented which turned liquid oils into solid fat that was inexpensive, suitable for frying and baking, and didn't go rancid. Since then, manufacturers replaced healthy oils with hydrogenated fats in thousands of products, until they now comprise about 8 percent of daily calories. These oils are detrimental to our health and have been implicated in cancer, heart disease and inflammatory conditions. At the same time, many of us are deficient in essential fatty acids, especially the Omega 3 and Omega 6 fatty acids that are in seafoods, grains, nuts, seeds and flaxseeds. These essential fatty acids are critical for growth, healing, reduction of pain and inflammation, healthy skin, reproduction, nervous system functioning, and overall well-being.

There have been significant reductions in the incidence of heart disease since 1970. Most researchers attribute this to a change in diet, plus an increase in exercise. An interesting controversy has sprung up around dietary cholesterol, which is often blamed for causing heart disease. Reducing the amount of cholesterol you eat is only a small part of the story. We eat no more cholesterol now than we did 80 years ago, yet heart disease has risen. In fact, only 30 percent of all people who eat excessive amounts of cholesterol have a change in serum cholesterol levels. Tests show a primary cause of high serum cholesterol is poor liver function.

Many researchers believe that an increase in vitamin C consumption may be the reason for the decline in cardiovascular disease. In countries where vitamin C intake has risen, heart disease has fallen. In other countries, cardiovascular disease is still on the rise. Oxidation of cholesterol is what causes problems. Oxidation, caused by free radicals in our food and environment, nicks the arterial lining, causing low-density-cholesterol (LDL) to stick to it. Calcium binds to the cholesterol, more LDL's sticks to the rough surface, more calcium bonds, and the process keeps repeating. The net result is plaque build-up in arteries. Vitamin C keeps LDL cholesterol from oxidizing, helps normalize serum cholesterol levels, and reduces atherosclerosis.[3] Dr. Matthias Rath, Director of Cardiology Research at the Linus Pauling Institute, discovered in 1989 that lipoprotein A exists in abnormal amounts in people with heart disease: "There is a direct relationship between vitamin C and lipoprotein A which leads to the conclusion that vitamin C deficiency is the primary cause of heart disease."[4] Another recent study concluded that men who took 300 mg or more of vitamin C daily had a 30 percent reduction in mortality from all causes.[5]

The average person consumes 12 grams of fiber daily, according to studies done by the USDA and the National Institutes of Health. This falls far short of the recommendation of 20 to 30 grams and is half of what people ate 150 years ago. Dietary fiber, found in fruits, vegetables, legumes and whole grains, is beneficial to our digestive tract and reduces risk of GI illness. In fact, short-chained fatty acids (like butyrate, the main fuel of the colon) are manufactured directly from fiber-rich foods and protect us against diseases of the colon.

Not only what we eat, but the way we eat has changed—for the worse. Often, we eat the same way we put gas in our cars: stop, fuel, go. We eat 45 percent of meals away from home, up from 39 percent in 1980 and 34 percent in 1980. Many of us skip breakfast, and others skip breakfast and lunch. Studies show that school-aged children perform better when they've eaten breakfast. Adults are no different. In fact, small, frequent meals keep our energy levels even and our minds alert. We need to treat our bodies like a wood stove.

You light the stove in the morning so it will warm the house while you work. Throughout the day you put small amounts of wood into the stove, and at nighttime, you fuel the stove and then let it die down before going to sleep. Our bodies work the same way. If we eat small meals and nutritious snacks throughout the day, they are burned efficiently.

Americans often overeat socially and emotionally. How commonly we hear that someone went to a party or out to dinner and just "stuffed themselves sick" or "ate like a pig." At home we eat when we aren't hungry because we are bored, lonely, angry, depressed, sad, frustrated, procrastinating and a million other reasons that have nothing to do with hunger or nourishing our bodies. We stuff feeling inside with food rather than feel the hurt inside. This too contributes to digestive illness.

We eat to give nourishment to our bodies, but meals are also a time for relaxation, rest, refreshment and renewal. If we are relaxed while eating, we digest food better. Numerous studies show that our mental state while eating has bearing on how well we digest food. People seem to know this intuitively. Saying grace or taking a couple of moments to center ourselves before eating is a global custom.

FOOD AND THE ENVIRONMENT

Thousands of years ago, people foraged and hunted for food. When populations increased, people learned how to farm and propagate plants and animals, so that more people could be fed with regularity. Farmers used to grow many foods, rotate crops and use "natural" fertilizers. Foods grown nearby were eaten fresh and primarily in season.

Today, we depend on the global economy to produce and supply our food. It seems perfectly normal to buy Granny Smith apples from New Zealand, dried Turkish apricots, salmon from Norway, and drink mineral water and wine imported from France or California. We have grown accustomed to nectarines in winter and oranges in summer.

Are foods shipped from far away just as nutritious as foods that are grown locally? A ripe, juicy tomato from your backyard has about the same measurable nutritional value as those whitish-orange hothouse tomatoes on sale in winter. But even though they have the same "scientific" measurements, our intuitive measurement tells us they are different.

The life force in foods gives us vitality and life. If you photograph foods with Kirlian photography, living foods have large energy fields, while processed foods have little or none. Once a plant is picked or an animal killed, a grain split or milk homogenized, it begins to lose its essential life force. Transporting foods over long distances diminishes their life-giving capacity. Statistically, canned, frozen and packaged foods often contain great nutrients, but we know that they're different from fresh or homemade foods. In fact, processed foods are enzyme-deficient. They don't have the essential enzymes that are critical aids in digestion and metabolism. Fresh fruits, vegetables, local fish and game, grains, beans, nuts and seeds give us these necessary enzymes. If your body doesn't have to work overtime making enzymes, it has more energy for other processes. Whole foods are in balance with themselves and with nature. When we eat them, we benefit from their balance.

Today our soils are being depleted. Most food in America is grown on corporate agrifarms that grow monocrops. Chemical fertilizers add only the nutrients necessary for healthy plants, not nutrient-rich foods. In fact, use of pesticides and herbicides to limit crop loss has skyrocketed over the past 30 years. The average person consumes 1 pound of these chemicals each year. These pesticides have neurotoxic effects and can cause damage to our nervous systems. They are especially harmful to children whose small bodies are exposed to more pesticides per unit of weight than adults.

The good news is that organic farming and integrated pest management are gaining momentum. The USDA has taken an active role in setting up standards for organic growers.

Most of our produce is hybridized. Its nutrient value is often sacrificed for pesticide resistance, ease of transportation or appearance. Many of these hybridized foods look or taste better than their old counterparts, but corn, for instance, has 14 percent less protein now than it did forty years ago.

In Hawaii tiny wild bananas are much more flavorful than commercial bananas. Throughout the world groups of people now collect seeds from these nonhybrid food plants and grow them. Someday we may be very grateful for these pioneers who are helping to protect biodiversity.

FOOD PREPARATION AND TECHNOLOGY

Linked to changes in food production and distribution are changes in food preparation. Valuable nutrients are taken out of foods, while chemicals and processes that change the very genetic makeup of foods are used to preserve and "enhance" them.

Food processing destroys nutrients. For example, whole wheat contains 22 vitamins and minerals that are removed to make white flour. After the bran and germ are removed from the whole wheat kernel, so too are 98 percent of pyridoxine (vitamin B6), 91 percent of manganese, 84 percent of magnesium and 87 percent of fiber. One of the many lost nutrients is chromium, which is critical for maintenance of blood sugar levels, normalizing high serum cholesterol levels and fat burning. The average American diet is seriously depleted in chromium due to food processing.

From the earliest times, people salted meats and other foods to cure them. Later they canned foods with sugar, salt and vinegar to keep them from perishing. Today, because food is produced and shipped from afar, manufactured chemical additives are put into foods to stabilize and preserve them. More than three thousand food additives are used in the United States alone—dyes, artificial flavors, dough conditioners, texturing agents, anti-caking agents, and so on—to extend shelf life and enhance flavor, appearance, consistency and texture. The average person eats an alarming 3 to 12 pounds of additives each year.

Current research indicates that only a tiny percentage of the population is sensitive to food additives, but I have seen many people in my practice with sensitivities. One 15-year-old client went into anaphylactic shock within 10 minutes if she ate a single drop of food coloring. Another client could no longer eat in restaurants because of severe reactions to monosodium glutamate. It is well documented that sulfites cause asthma and respiratory problems in sensitive individuals.

The long-term effects of food additives on children are of special concern. They consume more harmful substances per body weight than do adults. Ben Feingold, among others, found that many additives caused significant behavior and learning problems in sensitive children. The potential for long-term damage is clearly greater for children than for adults.[6]

Additives have been tested singly, but never in combination. What are the long-term effects? No one really knows. The chemistry experiment going on inside of us reminds me of my favorite experiment with a childhood chemistry set. Whenever I mixed the chemicals together, the test tube would explode. Though healthy people can handle most food additives, why burden your body with having to detoxify them? Minimize them by reading labels carefully.

Americans love the convenience of frozen foods. But if you read the labels, you'll find that most frozen foods contain additives that make the foods less perishable or less expensive. With careful shopping, you can find good quality frozen foods.

Microwave cooking has spread like brush fire over the last

two decades. Yet there is no scientific consensus about the safety of microwaved food. Many researchers find that although there are some particles in microwaved food that aren't ordinarily found in nature, the food remains nutritious and healthful. Others feel that microwaved food is denatured and therefore detrimental to our health. Studies show that when breast milk has been microwaved to 98.6°F, almost all the antibodies and lysozymes that protect us from infection are destroyed.[7] Still other studies are conflicting. One thing is for sure: microwaving foods is too recent an innovation to know what the long-term effects are. It really doesn't take that much longer to cook food in the oven or on the stove.

For the first time, we now have genetically engineered foods. Technologists are splicing genes into dozens of foods to make them last longer, be juicer, taste sweeter, grow bigger, and be more pest-resistant. Some experts are opposed to bioengineering. They are concerned that people with allergies to specific foods will react to gene-spliced foods. What will these foods do to the long-term ecological balance of the world?

Food irradiation is now being used in food production. It's a "clever" way to use nuclear wastes—to keep food fresh longer and decrease incidence of food poisoning. Irradiation kills all bacteria, like Salmonella, and leaves no radiation in the food itself. But many researchers are opposed to irradiation. For instance, milk loses 70 percent of vitamin A, thiamin and riboflavin. Moreover, irradiated foods have molecules that are found nowhere in nature. The Food and Drug Administration (FDA) dubs them "radiolytic byproducts" and separates them into two categories: "known radiolytic products," such as formaldehyde and benzene, which are known carcinogens, and "unique radiolytic products," which are new molecules that haven't been characterized. What's frightening is no one knows what the long-term effects of these molecules will be on health. Many people are opposed to such massive experimentation done at our risk. They are also worried about the risk of having small irradiating facilities throughout the country, which will have all the problems associated with handling nuclear materials.

The Changing American Lifestyle

Though many people have a high standard of living today, the price is a hurried life that takes its toll on our bodies. Studies show we actually have less leisure time than people had just 20 years ago—despite an array of time-saving appliances. Because our bodies and our minds work together, the stresses we feel in either one affects the other. The synergy can help calm us down or stress us out. The mind/body connection plays an important role in digestive wellness. Stress makes our stomach feel like it has rocks in it. Stress plays a large role in ulcerative colitis, skin conditions and autoimmune problems. In fact, all health problems are due to stress: physical, emotional, or environmental.

Commonly used drugs also play a role in the development of many digestive illnesses. Seventy million prescriptions for nonsteroidal anti-inflammatory drugs (NSAIDs) are written each year. The cost this contributes to the development of gastric ulcers is estimated to be over $100 million a year. Aspirin is especially hard on the stomach lining and overuse can contribute to the development of ulcers. NSAIDs, like Tylenol, Motrin, Advil and dozens of others, are gentler on the stomach lining but much more irritating to the intestinal lining. They cause damage to the lining by blocking prostaglandins that stimulate repair. They are a direct cause of leaky gut syndrome, food sensitivities and inflammatory problems like arthritis and eczema.

Antibiotics kill not only disease-causing bacteria, but healthy ones as well. Healthful bacteria, like *Lactobacillus acidophillus*, attach tightly to the intestinal lining so that no parasites or disease-producing organisms can get a foothold.

Antibiotics kill these friendly bacteria, allowing pathogenic microbes, viruses and fungi to take hold. Antibiotics also disrupt the natural symbiosis of the gut and can cause gross imbalance of the natural flora, leading to chronic and systemic illness. According to S. M. Wolfe, "After Congressional hearings and numerous academic studies on this issue, it has become the general consensus that 40-60 percent of all antibiotics in this country are misprescribed."[8]

We live in a toxic environment. Air quality is questionable, and water, our most precious resource, is becoming polluted. Although various localities are cleaning up their natural resources, the global balance is on average deteriorating. Our soils are becoming contaminated. Vast numbers of grazing livestock are destroying many habitats and causing erosion. World population has exploded, demanding material needs that are stripping the planet.

Many physicians believe that the underlying cause of digestive illness is a combination of poor nutrition and exposure to toxic substances. Not only are we exposed daily to hundreds of chemicals—side-stream smoke, chlorine and fluoride from water, air pollution, cosmetics, toiletries, household cleaning supplies, medications and workplace toxins. Healthy people can handle them fairly well, but when the liver gets overloaded, it fails to adequately protect us. Many holistic therapies integrate detoxification programs, which enhance the liver's ability to carry us through this chemical minefield. If our livers are functioning suboptimally, our digestion doesn't work optimally, which leads to increased work for the already overtaxed liver, which leads to more digestive problems, and on and on.

We have yet to find the balance between the miracles that technology offers and the health of our environment. Our greatest challenge is to increase the standard of living in a way that everyone on earth can reap the many benefits of modern technology without destroying our ecosystems and biodiversity.

The Will to Change

Most people are unaware of how closely their health problems are related to their lifestyle. We are overfed and undernourished. Poor food choices, use of alcohol and cigarettes, a sedentary lifestyle and chronic stress shorten our lives and contribute to degenerative illness as well as how we feel daily. People are born with reserves of nutrients in their organs, but a lifetime of low-nutrient foods will deplete those reserves, weakening the body's ability to heal. We like to think of ourselves living robustly, rooting and tooting until we die, but the truth is most of us limp along, accepting poor health and declining quality of life for the last 20 years of our lives as though it were normal and customary. Many of us have come to accept digestive illness as an integral part of our lives. But we can change the way we feel by making changes in our way of life.

It's time to focus on health rather than convenience and develop better habits. Instead of asking, Does it look and taste good?, we should ask: Is this food healthful, will it contribute to my biochemical balance and help me feel better, and will it taste good? We need to exercise regularly and think positive thoughts. We need to relax by ourselves and with friends. We need to create balance in our lives.

Your digestive problems offer you an opportunity to change. You can see this as a curse or a blessing. Donald Ardel, father of Wellness, calls it a "teachable moment." You're so sick and tired of being sick and tired that you're willing to make some real changes! This involves commitment on your part. Remember: no pain, no gain!

Exercise 2

Food Diary

Write down everything you eat and drink in a diary and keep track of how your body feels. If you have diarrhea or pain after you eat or feel like a million bucks, write it down.

See if you can correlate specific foods to the way your digestive system works. Keep track of where you were, who you were with and your moods. Digestive problems are often related to how comfortable or uncomfortable we feel in a situation. Does a specific person make you nervous? Keeping a chart will help you find out. A sample chart looks like this:

DAY

Time of day	What did I eat?	Where was I? Who was I with? What was my mood?	How does my GI tract feel?

EXAMINING YOUR FOOD DIARY

1. Do you eat breakfast every day? Breakfast provides the fuel we need to get our bodies going for the day. It literally means "break the fast."
2. Is your indigestion better or worse at specific times of the day? This can be a clue to indigestion. Maybe it's

something you ate, how fast you ate or how much you ate.

3. Do you eat when you aren't really hungry? If so, examine the reasons, and try to find other outlets for your energies.
4. How often do you eat? Most people feel best when they eat three meals daily plus nutritious snacks. This meal plan keeps blood sugar levels even and facilitates digestion.
5. Do certain foods/beverages provoke symptoms? Eliminate suspicious foods for a week and note any differences in how you feel.
6. Are you relaxing at mealtimes or rushing? Eating more slowly aids digestion.
7. Do you get at least five servings of fruits and vegetables each day? A serving is a piece of fruit, a half cup of most vegetables or a cup of lettuce. Five servings is a minimal requirement.
8. What percentage of your foods and beverages were high-sugar, high-fat, low-fiber or highly processed? Replace these with fresh, wholesome foods.
9. Do you consume enough high-fiber foods? We find fiber in whole grains, fruits, vegetables and legumes.
10. Do you drink enough—6 to 8 cups—water, herb teas and juices? Soft drinks and coffee don't count!

Part 2

Healthy Digestion/ Faulty Digestion

Chapter 3

A Voyage Through the Digestive System

"The surface area of the digestive mucosae, measuring up and down and around all the folds, rugae, villi and microvilli, is about the size of a tennis court."

SYDNEY BAKER, M.D.

THE DIGESTIVE SYSTEM is self-running and self-healing. Because this beautiful, intricate system works automatically, the average person knows very little about it. Let's take a trip through the digestive system to see what miraculous events occur inside us every moment of our lives.

Think of the digestive tract as a 25- to 35-foot hose that runs from the mouth to the anus. Its function is to turn the foods we eat into microscopic particles that the cells can use for energy, maintenance, growth and repair. The old saying "You are what you eat" is primarily true. From birth to death, we continually create and recreate ourselves from the nourishment we put inside the body. But not everything we ingest is good. Fortunately, the digestive system keeps toxic substances from entering the bloodstream. The liver lets hydrogenated oils, chemicals, pesticides, herbicides and drug components pass through, breaks them down, excretes them or stores them in the liver. The process is very selective, having evolved over time to work optimally. Unfortunately, the digestive system has had difficulty adapting to the explosion of environmental, technological and social changes humans have created over the past few decades. So it's not only what you eat, but what you digest that's important.

Whatever we eat is squeezed through the digestive system by a rhythmic muscular contraction called peristalsis. Sets of smooth muscles contract alternately, pushing food through the esophagus to the stomach and through the intestines. There it is acidified, liquefied, neutralized and homogenized until it's broken down into usable particles. From the time you swallow, this process is involuntary and can occur even if you stand on your head. My seven-year-old son demonstrated this by eating upside-down. Yes, the food went down—or rather up—as usual!

Along the journey, the body breaks food down—protein into amino acids, starches into glucose and fats into fatty acids and glycerol. Enzymes, vitamins and minerals are also absorbed. The cells use these raw materials for energy, growth and repair. When the digestive system doesn't work, individual cells gradually lose their ability to function properly. This is especially apparent in the lining of the digestive tract which repairs and replaces itself every three to five days.

YOU AREN'T ONLY WHAT YOU EAT

Nutritious foods are the right place to start on the path to digestive wellness. But many people eat all the "right" foods and still have digestive problems. The best diet in the world won't help malabsorption problems. You must

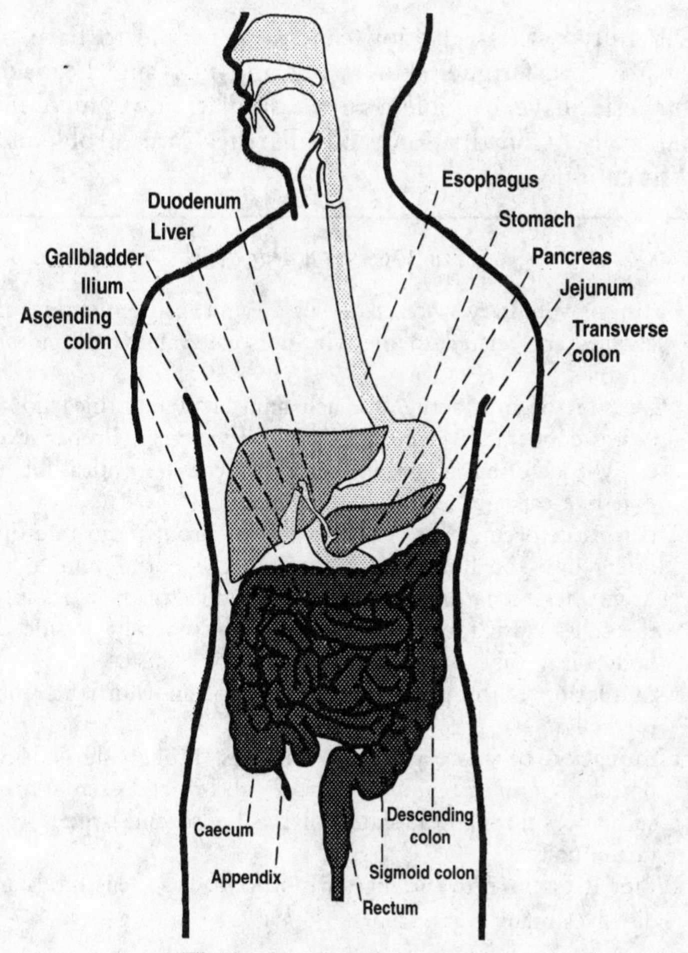

The human digestive system.

be able to digest foods; break them down into tiny particles; absorb the food mash; take that through the intestinal lining and into the bloodstream; assimilate nutrients and calories; take them into the cells where they can be used; and eliminate waste products through the kidneys, bowels, lymph system and skin. Health can and does break down at any of these phases. For example, people with lots of intestinal gas are fermenting their food.

Difficulty with absorption can cause people to have food sensitivities, fatigue, skin rashes and migraine headaches. Diabetics have a problem with assimilation of glucose into the cells. Constipation and diarrhea are problems of elimination.

THE DIGESTIVE PROCESS

Eating is voluntary when materials are put in the mouth. Food choices are related to lifestyle, personal values and cultural customs.

Digestion occurs in the stomach and small intestine and requires cooperation from the liver and pancreas. Proper levels of hydrochloric acid and intestinal bacteria are critical for full digestive capacity.

Absorption occurs when food is taken through the intestinal lining into the bloodstream through the portal vein to the liver where it is filtered. From the bloodstream it passes to the cells. Until food is absorbed, it is essentially outside the body—in a tube going through it.

Assimilation is the process by which fuel and nutrients enter the cells.

Elimination of waste products happens through the kidneys, bowels, lymph system and skin. We absorb and excrete many substances through the skin, which is the second largest organ in the body.

Water is essential to the digestive process. It softens foods and dissolves many components.

Digestive Process	Where in the Body	Function
Eating/food choices	Mouth/mind	Portal for all nutrients/materials to enter the body
Digestion	Stomach/small intestine; to a lesser degree, saliva in the mouth	Breaks food down into basic components for use by cells

Digestive Process	Where in the Body	Function
Absorption	Small intestines/ large intestines, bloodstream, liver	Food comes through intestinal wall into bloodstream
Assimilation	Cellular	Nutrients enter cells & are used for energy, storage & structure
Elimination	Colon, kidneys, skin, lymph system, cells, blood stream	Wastes are excreted

A Guided Tour Through the Digestive System

To gain a thorough understanding of how the digestive system works, let's take a guided tour starting at the mouth and ending at the colon.

Mouth

The main function of the mouth is chewing and liquefying food. The salivary glands, which are located under the tongue, produce saliva which softens food, begins dissolving soluble components and helps keep the mouth and teeth clean. Saliva contains amylase, an enzyme for splitting carbohydrates. Only a small percentage of starches are digested by amylase, but if you keep a piece of bread in your mouth for a long time, you can begin to taste the increased sweetness that comes from splitting the starch into simple sugars. Chewing also stimulates the parotid glands, behind the ears in the jaw, to release hor-

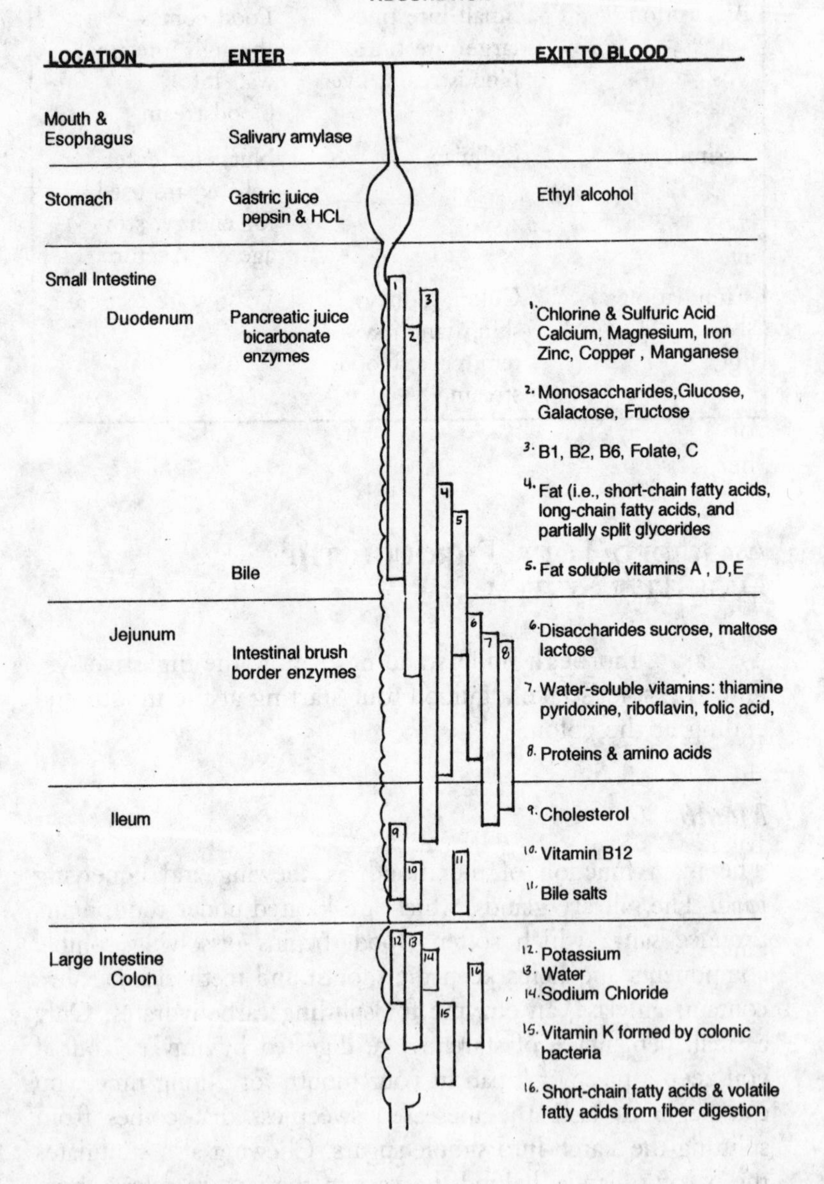
DIGESTIVE SYSTEM
ABSORBTION
LOCATION
ENTER
EXIT TO BLOOD
Mouth & Esophagus
Salivary amylase
Stomach
Gastric juice pepsin & HCL
Ethyl alcohol
Small Intestine
Duodenum
Pancreatic juice bicarbonate enzymes
Bile
Jejunum
Intestinal brush border enzymes
Ileum
Large Intestine Colon
1. Chlorine & Sulfuric Acid Calcium, Magnesium, Iron Zinc, Copper , Manganese
2. Monosaccharides, Glucose, Galactose, Fructose
3. B1, B2, B6, Folate, C
4. Fat (i.e., short-chain fatty acids, long-chain fatty acids, and partially split glycerides
5. Fat soluble vitamins A , D,E
6. Disaccharides sucrose, maltose lactose
7. Water-soluble vitamins: thiamine pyridoxine, riboflavin, folic acid,
8. Proteins & amino acids
9. Cholesterol
10. Vitamin B12
11. Bile salts
12. Potassium
13. Water
14. Sodium Chloride
15. Vitamin K formed by colonic bacteria
16. Short-chain fatty acids & volatile fatty acids from fiber digestion

mones that stimulate the thymus to produce T-cells, which are the core of the protective immune system.[1]

Healthy teeth and gums are critical for proper digestion. Many people eat so fast, they barely chew their food at all and then wash it down with liquids. So the stomach receives chunks of food instead of mush. This undermines the function of the teeth, which is to increase the surface area of the food. These people often complain of indigestion or gas. In *May All Be Fed*, John Robbins describes three men who survived in a concentration camp during World War II by chewing their food very well, while their compatriots perished. Simply by chewing food thoroughly we can enhance digestion and eliminate some problems of indigestion.

The most common problems that occur in the mouth are sores on the lips or tongue—usually canker or cold sores (herpes)—and tooth and gum problems.

Esophagus

The esophagus is the tube which passes from the mouth to the stomach. Here peristalsis begins to push the food along the digestive tract. Well-chewed food passes through the esophagus in about 6 seconds, but dry food can get stuck and take minutes to pass. At the bottom of the esophagus is a little door called the cardiac or esophageal sphincter. It separates the esophagus from the stomach, keeping stomach acid and food from coming back up. It remains closed most of the time, opening when a peristaltic wave, triggered by swallowing, relaxes the sphincter. The most common esophageal problems are heartburn (also called gastric reflux) and hiatal hernia.

Stomach: The Body's Blender

The stomach is the body's blender. It chops, dices and liquefies as it changes food into a soupy liquid called chyme, which is the beginning of the process of protein digestion. The stomach is located under the rib cage, just under the heart.

Protein molecules are composed of chains of amino acids—up to 200 amino acids strung together. Hydrochloric acid (HCl), produced by millions of parietal cells in the stomach lining, begins to break these protein chains apart. HCl also kills microbes that come in with food, effectively sterilizing it. HCl is so strong that it would burn our skin and clothing if spilled on us. Yet, the stomach is protected by a thick coating of mucus (mucopolysaccharides), which keeps the acid from burning through the stomach lining. Prostaglandins, small chemical messengers, help keep the mucous layer active by sending messages to replace and repair the stomach lining and provide a protective coating. When this mucous layer breaks down, HCl burns a hole in the stomach lining, causing a gastric ulcer.

The stomach also makes pepsin, a protein-splitting enzyme, which cuts the bonds between specific amino acids, breaking them down into short chains of just 4 to 12. The stomach also produces small amounts of lipase, enzymes that digest fat. Most foods are digested and absorbed farther down the gastrointestinal tract, but alcohol, water and certain salts are absorbed directly from the stomach into the bloodstream. That's why we feel the effects of alcohol so quickly.

VITAMIN B12 AND INTRINSIC FACTOR

Before vitamin B12 even had a name, scientists knew that there was something in food that joined with something in the stomach that helped its absorption. They named these two substances intrinsic factor and extrinsic factor. Later, extrinsic factor was discovered to be vitamin B12, which is essential for blood formation, energy, growth, and cell division and function. Intrinsic factor is made in the stomach and binds vitamin B12 so that it can be readily absorbed in the intestines. Adequate hydrochloric acid is required for the production of intrinsic factor.

As we age, the ability to manufacture hydrochloric acid decreases. Intrinsic factor is likewise decreased and vitamin B12 deficiencies occur. Many elderly people have vitamin B12 deficiencies which affect the body's ability to get oxygen into each cell. The main symptom is muscle weakness and fatigue. Many people benefit from vitamin B12 injections, under a physician's care, even though many do not have low serum B12 levels or pernicious anemia (anemia caused by B12 deficiency). I remember one elderly woman who had normal serum B12 levels, but she felt enormously different when B12 shots were added to her regimen. This simple, inexpensive therapy can dramatically affect quality of life for those who need it. More sensitive tests of vitamin B12 status are also available—homo-cysteine and methylmalonic acid.

Food stays in the stomach 2 to 4 hours—less with a low-fat meal, more with a high-fat or high-fiber meal. When the stomach has finished its job, chyme has the consistency of split pea soup. Over several hours it passes in small amounts through the pyloric valve into the duodenum, the first 12 inches of the small intestine. Chronic stress lengthens the amount of time that food stays in the stomach, while short-term stress usually shortens the emptying time. Most of us have experienced a nervous stomach or a feeling in the pit of the stomach or a stomach that feels like it's filled with rocks.

The most common problems associated with the stomach are gastric ulcers and underproduction of hydrochloric acid.

The Man with a Window in his Stomach

We owe early information about how digestion works to a man who had a "window" in his stomach. In 1822, Alexis St. Martin, a young Canadian trapper, was shot at close range in his left side. A U.S. Army doctor, William Beaumont, successfully treated St. Martin but was unable to close the wound completely. Over the next eight years Dr. Beaumont observed the activity of St. Martin's stomach through the hole, watching what happened when various foods were eaten and how his emotions affected his digestive activity. Until this time, people had no idea what happened to food after it was swallowed. Science owes much to these two men.

Small Intestine

The small intestine is hardly small. If this coiled-up garden hose were stretched out, it would average 15 to 20 feet long. If spread flat, it would cover a surface the size of a tennis court. Here food is completely digested and absorbed. Nutrients are absorbed through hundreds of small fingerlike folds called villi in the intestinal wall, which are covered, in turn, by millions of microvilli. (Think of them as small loops on a velvety towel which then have smaller threads projecting from them.) The villi and microvilli are only one layer thick, but perform multiple functions of producing digestive enzymes, absorbing nutrients and blocking absorption of substances that aren't useful to the body.

The intestinal lining repairs and replaces itself every three to five days. The sloughed material contains enzymes and fluids which are recycled to help digestion. The intestinal wall has a paradoxical function: it allows nutrients to pass into the bloodstream while blocking the absorption of foreign substances found in chemicals, bacterial products and other large molecules found in food. Some foods we eat and medications we use cause the intestinal wall to lose the ability to discern between nutrients and foreign substances. When this occurs, there is a problem of intestinal permeability, or leaky gut syndrome. This newly recognized syndrome contributes to skin problems, food sensitivities, osteoarthritis, migraine headaches and chronic fatigue syndrome.

The small intestine has three parts: the duodenum, the jejunum and the ileum. The duodenum is the first 12 inches of the small intestine, the jejunum is the next 40 percent, and the ileum is the last segment. Each nutrient is absorbed at specific parts of the small intestine. For instance, the duodenum has an acidic environment which facilitates absorption of some nutrients, including calcium, copper, iron, folic acid, thiamin, manganese, vitamins A and B2 and zinc. People with low hydrochloric acid levels may become deficient in one or more of these nutrients.

RELATIVE IMPORTANCE OF SITE ABSORBTION
WITHIN THE SMALL INTESTINE

LOCATION	LENGTH	EXITS TO BLOOD
Duodenum	12-18"	Chlorine , Sulfuric Acid Calcium, Magnesium, Iron Zinc, Copper, Manganese
		Monosaccharides, Glucose Galactose, Fructose
		B1, B2, B6, Folate, C
		Fat (i.e., short-chain fatty acids, long-chain fatty acids, & partially split glycerides)
		Fat soluble vitamins A, D, E, H
Jejunum	10'	Disaccharides sucrose, maltose, lactose
		Water-soluble vitamins: thiamine, pyridoxine, riboflavin, folic acid
		Proteins & amino acids
Ileum	12'	Cholesterol
		Vitamin B12
		Bile salts

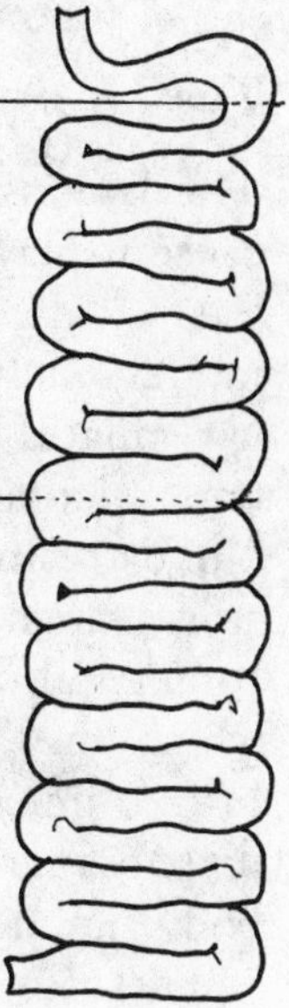

Pancreas

The pancreas has two main roles: production of digestive enzymes and insulin. When food passes from the stomach to the duodenum, the pancreas secretes bicarbonates, essentially baking soda, which neutralizes the acidity of the chyme so it won't burn the tissues of the intestines. Then it manufactures and secretes specific enzymes that digest fats, carbohydrates and protein. These enzymes are lumped into three categories: lipase, amylase and protease. Lipase breaks fats into fatty acids and glycerol; amylase splits carbohydrates into simple sugars; and protease digests the links between amino acids from protein. Once digested, these nutrients can be absorbed into the bloodstream and used by the cells.

The second role of the pancreas is regulation of blood sugar levels. When the blood sugar is too high, the pancreas secretes insulin which signals the cells to store glucose. When this mechanism fails, people develop diabetes.

The Liver: The Body's Fuel Filter

The liver is the most overworked organ in the body because it plays many roles. It manufactures bile to emulsify fats for digestion; it makes and breaks down many hormones, including cholesterol, testosterone and estrogens; it regulates blood sugar levels; it processes all food, nutrients, alcohol, drugs and other materials that enter the bloodstream and lets them pass, breaks them down or stores them. In our culture, the job of neutralizing environmental pollution inside the body is no small task—one which the liver never evolved to handle. The liver can lose as much as 70 percent of its capability and not show diagnosable liver disease. The quality of virtually every body function depends on the liver.

The four and one-half pound liver manufactures 13,000 chemicals and has 2,000 enzyme systems, plus thousands of synergists that help with body functions.[2] With these chemicals and enzymes, it "humanizes" nutrients so that the cells can use them. Practically all vitamins and minerals we take in need to be enzymatically processed by the liver before we can use them. If the liver is too congested to enzymatically process these nutrients, we do not get the benefit from them.

Bile, manufactured by the liver and stored by the gallbladder, buffers the intestinal contents due to its high concentration of bicarbonates. It also emulsifies fats. Bile is a soaplike substance that makes fats more water-soluble, increasing their surface area so that the enzymes can split them for the cells to use.

The liver has three lobes: main, left and lower. The main lobe organizes and humanizes nutrients. It is the main chemical factory, producing enzymes and chemicals necessary for body functions. The left lobe regulates and maintains body functions. People with toxic left lobes are often environmentally sensitive or pan-allergic—allergic to nearly everything. Many times allergic people crave what they are sensitive to. The body gets used to having nicotine, alcohol, wheat, dairy or whatever, and when we remove it, the body's balance is disturbed. The left lobe works to maintain body homeostasis (staying the same) without the "missing" substance. The craving is, in some part, the liver's way of trying to get us what we "need." As Jack Tips writes in *The Liver Triad*, "As long as toxic residues from these substances are present in this lobe, the body will get a subtle signal to continue the addiction, to want to respond to allergens."[3]

The lower lobe is where the essential fatty acids and fat-soluble vitamins A, D, E and K are stored so the liver and other glands can produce cholesterol and hormones. This is also where the liver stores environmental toxins like radioactive substances, pesticides, herbicides, food preservatives and dyes. The liver will detoxify what it can, but if it can't break down a particular substance, it stores it. It is important to detox the liver on a regular basis, perhaps twice a year, to help maintain its function. Many systems have been developed to help detoxify the liver, which we discuss at length in Chapter 9.

Gallbladder: A Holding Tank for Bile

The gallbladder is a pear-shaped organ that lies just below the liver. The gallbladder's function is to store and concentrate bile, which is produced by the liver. Bile emulsifies fats, cholesterol and fat-soluble vitamins by breaking them into tiny globules. These create a greater surface area for the fat-splitting enzymes (lipase) to act on during digestion. When we eat, the gallbladder and liver release bile into the common duct which connects the liver, gallbladder and pancreas to the duodenum. Between meals the gallbladder concentrates bile. The most common problem of the gallbladder is gallstones. When bile becomes too concentrated, stones may form, which can cause pain and discomfort. Gallbladder disease is directly related to diet.

The Large Intestine or Colon

When all nutrients have been absorbed, water, bacteria and fiber pass through the ileocecal valve to the large intestine or colon. The ileocecal valve is located by your right hipbone and separates the contents of the small and large intestine. The appendix is a small, fingerlike sac which extends off the beginning of the colon. Until recently, the function of the appendix was a mystery. Now we know it contains a great deal of lymphatic tissue and is thought to be part of the immune system.

The colon is short, only 3 to 5 feet long. Its job is to absorb water and remaining nutrients from the chyme and form stool. Two and a half gallons of water pass through the colon each day, two-thirds of which come from body fluids. The efficient colon pulls 80 percent of the water out of the chyme, which is absorbed into the bloodstream.

The large intestine has three main parts: the ascending colon (up the right side of the body), the transverse colon (straight across the belly under the ribs) and the descending colon (down the left side of the body) to the rectum, where feces exit the body. Stool begins to form in the transverse colon. If the chyme passes through the colon too quickly, water is not absorbed, causing diarrhea. Stool that sits too long in the colon becomes dry and hard to pass, leading to constipation. About two-thirds of stool is composed of water and undigested fiber and food products. The other third is composed of living and dead bacteria.

The large intestine contains trillions of bacteria. Helpful bacteria, called probiotics, lower the pH of the colon, killing disease-causing microbes. Probiotics also produce vitamins B and K, protect us from illness, enhance peristalsis and make lactase for milk digestion. Probiotic bacteria ferment the dietary fiber, producing short-chained fatty acids—butyric, propionic, acetic and valerate. Most of the tissues of the body prefer to use glucose as a fuel, but the colon prefers to burn butyric acid. Low butyric acid levels or an inability of the colon bacteria to properly metabolize butyric acid has been associated with ulcerative colitis, colon cancer, active colitis and inflammatory bowel disease.

When the stool is finally well formed, it gets pushed down into the descending colon and then into the rectum. It is held there until there is sufficient volume to have a bowel movement. Two sphincters—rings of muscle—control bowel movements. When enough feces have collected, the internal spincter relaxes and your mind gets the signal that it's time to relieve yourself. The external sphincter opens when you command it. Because this is voluntary, you can have the urge to defecate, but wait until it's convenient. If you ignore the urge, water keeps being absorbed back into the body and the stool gets dry and hard. Some people are chronically constipated because they don't want to take the time to have a bowel movement or don't like to have bowel movements at work. This book is about listening to your body signals. Take the time when your body calls you, not when it's convenient or ideal.

Many health problems arise in the colon: appendicitis, constipation, diarrhea, diverticular disease, Crohn's disease, ulcerative colitis, rectal polyps, colon cancer, irritable bowel syndrome, parasites, hemorrhoids.

WHAT GOES IN, MUST COME OUT!

We can learn a lot about ourselves from stool. Dennis Burkitt, M.D., father of the fiber theory, found that on average people on Western diets only excreted 5 oz of stool daily, whereas Africans eating traditional diets passed 16 oz. Well-formed stool tells us when it wants to come out; we don't need to coax it. It looks like a brown banana with a point at one end, is well-hydrated and just slips out easily. Stool that looks like little balls all wadded together has been in the colon too long. The longer waste materials sit in the colon, the more concentrated the bile acids become; concentrated bile acids irritate the lining of the colon.

Frequency of bowel movements is a good health indicator. How often do you have a bowel movement? People on good diets generally have 1 to 2 each day. If you are not having a daily bowel movement, there can be many causes. First, take a close look at your diet. You probably aren't eating enough fiber. If not, increase your intake of fruits, vegetables, whole grains and legumes. These foods are generally high in magnesium, which helps normalize peristalsis. Make sure that you are drinking enough fluids. Coffee and soft drinks don't count!

Another good indicator of your colon's health is your bowel transit time—how long it takes food to move from the first swallow until it exits the body. When your system is working well, the average amount of time is 18 to 36 hours. On average, Americans have a transit time that is way too long—48 to 96 hours—because we don't eat enough high-fiber foods or drink enough water. You can do a simple home test to determine your transit time, which gives you important information about the way your body works.

Testing Bowel Transit Time

Transit time is how long it takes from the time you eat a food until it comes out the other end. Buy charcoal tablets at a drug or health food store, and take 20 grains (or 1 g)—anywhere from 5 to 12 depending on how big they are. Note exactly when you took the charcoal. When you see darkened stool (charcoal will turn the stool black), calculate how many hours since you took the charcoal tablets. That is your transit time. You can also do the test with beets. Eating 3 or 4 whole beets will turn stool a deep garnet red.

The Results

Less than 12 hours: This usually indicates that you are not absorbing all the nutrients you should from your food. You may have malabsorption problems.

12 to 24 hours is optimal.

More than 24 hours: This indicates that wastes are sitting inside your colon too long. Poor transit time greatly increases the risk of colon disease. Substances that were supposed to be eliminated get absorbed back into the bloodstream, and they can interfere with and irritate your system. Take action now! Increase your fiber intake by eating more fruit, vegetables, whole grains and legumes. Drink lots of water every day. Get 30 minutes of exercise at least 3 times a week.

Exercise 3

CLEAN UP YOUR DIET!

Let's take a look at what foods you are eating and begin the process of cleaning up your diet. Take last week's diary. Get out some crayons or markers. You're going to color!

Circle the following foods:

Sugar, caffeine, alcohol, junk foods, fried foods, high-fat foods, pastries, donuts, chips, microwave popcorn, highly processed foods, soft drinks, diet soft drinks, diet foods	Circle them red
Dairy products: milk, cheese, yogurt, ice cream, frozen yogurt, ice milk	Circle them blue
Fruits and vegetables	Circle them green
Protein foods: fish, poultry, beef, pork, lamb, veal, legumes, soy products	Circle them yellow
Nuts and seeds, oils, butter, margarine	Circle them purple
Grains: wheat, bread, corn, rice, millet, buckwheat, bulghur, quinoa, amaranth, barley, oats, rye	Circle them black

Look at those circles. Is there one food group that dominates your diary? If you eliminated one of these categories from your diet, which would be the easiest to give up and which would be the most difficult? Sometimes the ones which are the hardest to give up are the ones that are causing us the most trouble. They temporarily make us feel better, even though they are really making us sicker. Why? Our bodies react negatively to cigarettes, dairy products, caffeine, sugar, wheat, pork, beef, citrus fruits or any other foods, yet we crave them.

Tackle the Sweet Tooth

This week, focus on the foods you circled in red and eliminate all of them. Get rid of soft drinks, cookies, pastries, donuts, sugar added to coffee or tea. We're not talking about perfection here. Let's just make major progress. Why? Because these foods make it harder for your body to be healthy. High-sugar foods deplete our nutrient stores. We need most of the B-complex vitamins, chromium, manganese and potassium to metabolize these foods properly, but sweets don't have any of these nutrients. So we take nutrients out of storage, and eventually our tissues become depleted.

After a couple of weeks, fruit begins to taste really sweet, which is just how it ought to taste. Once, I realized that it had been months since I had had any chocolate. I began to feel deprived, so I bought a big chocolate bar for my family and friends. I ate a few squares and was totally satisfied. I hope that eventually you can be satisfied with just a little bit, too. But if you can't, you're really better off without any. Once I was sick and was craving sweets like crazy. My doctor told me it was the bacteria—both good and bad—that wanted the energy. So starve those bad guys out. The helpful bacteria can adapt with real food.

Chapter 4

The Bugs in Your Body!: Intestinal Flora

"There are more bacteria in our intestinal tract than cells in our body."

JEFFERY BLAND

WE HAVE 400 TO 500 types of bacteria in our digestive systems, each of which has many types of strains. This variety may seem overwhelming, but 20 types make up three-quarters of the total. The most common are bacteroides, bifidobacteria, eubacterium, fusobacteria, lactobacillus, peptococcaceae, rheumanococcus and streptococcus. Most of these bacteria are anaerobic, meaning they do not need oxygen to thrive; some are aerobic, which do need oxygen for survival; and a third group produces lactic acid and can be either aerobic or anaerobic. Lactic acid-producing bacteria help acidify the intestinal tract and protect us from overgrowth of harmful bacteria. A total of one hundred trillion bacteria live together in our digestive system, in either symbiotic or antagonistic relationships.[1] Their total weight is about 4 pounds—the size of the liver.

There are billions of bacteria in our mouths. While the stomach has few due to its high acid content, which prohibits their growth, the small intestine has many billions of bacteria. The overwhelming majority of intestinal flora reside in the colon—trillions and trillions. Each day we produce several ounces of these microbes and eliminate several ounces in stool. These bacteria manufacture substances that raise or lower our risk of disease and cancer, effect of drugs, immune competence, nutritional status and rate of aging. Some of these bacteria cause acute or chronic illness. Other bacteria cause illness in people who are genetically susceptible, but no problems in other people.

Another group of bacteria offer us protective and nutritive properties. These friendly bacteria are called "intestinal flora," "probiotics" or "eubiotics"—the last two terms mean "healthful to life." The term "probiotics" is commonly used to refer to supplemental use of these bacteria in powder or capsule form. The two most important groups of flora are the *lactobacilli*, found mainly in the small intestine, and *bifido bacterium*, found primarily in the colon. These bacteria live symbiotically within us in a mutually beneficial relationship that has evolved to enhance our health and theirs. They live at about 98.6°F and thrive on the constant nourishment we provide in a warm, dark, moist environment. We allow them to inhabit us because they give us valuable preventive and therapeutic benefits.

Main Bacteria Types in Our Bodies[2]

Type	Aerobic/ Anaerobic	Percent
Bacteroides, 20 species	Anaerobic	Almost 50
Bifidobacterium	Anaerobic	11
Pepto streptococcus	Anaerobic	8.9
Fusobacteria, 5 species		7
Rheumanococcus, 11 species		4.5
Lactobacillus	Both	2-2.5
Clostridia		0.6
Enterobacteria, *E. Coli*, Klebsiella, Aerobacteraerobacter, etc.		Less than 0.5

Where did these trillions of bacteria come from? Up until birth, we receive predigested food from our mothers and are born with a sterile digestive tract. The trip down the birth canal initiates us into the world of microbes that thrive everywhere. Babies are exposed to bacteria in breast milk and formula and when sucking on nipples, fingers and toes. With every breath and touch, bacteria enter the body to colonize on the skin and mucous membranes. In no time, every conceivable space in the colon is occupied by microbes. The microbes set up homogeneous neighborhoods which push out competing microbes that try to break into their territories. This normally happens in a predictable way, and once established, the colonies flourish. When babies are unable to properly colonize friendly flora, they become irritable, colicky, have gas pains and eczema in their diaper area. As we age, the specific types and strains of dominant flora change from *Bifidobacteria infantus* in infants to other strains of *Bifidobacteria* in children and adults.

THE MANY BENEFITS OF INTESTINAL FLORA

Flora play an important role in our ability to fight infectious disease, providing a front line in our immune defense. As noted in the 1988 report of the U.S. Surgeon General, "Normal microbial flora provide a passive mechanism to prevent infection." Friendly flora also manufacture many vitamins including the B-complex vitamins biotin, thiamin (B1), riboflavin (B2), niacin (B3), pantothenic acid (B5), pyridoxine (B6), cobalamine (B12) and folic acid, plus vitamin A and vitamin K.[3]

Lactic acid-secreting acidophilus and bifidus increase the bioavailability of minerals which require acid for absorption—calcium, copper, iron, magnesium, manganese. Farm animals are routinely given supplemental flora which enhance absorption of both vitamins and minerals.

Friendly flora help increase our resistance to food poisoning. In 1993 the United States reported 20 to 40 million cases of food poisoning, though the FDA estimates the true total to be 80 million. Many cases go unreported because the symptoms closely resemble the flu. Some food-borne infections lead to chronic illness, causing heart and valve problems, immune system disorders, joint disease and possibly even cancer. Use of supplemental acidophilus and bifidus can help prevent food poisoning by making the intestinal tract inhospitable to the invading microbes. It is a common misconception that friendly flora kill invading microbes. What they actually do is change the environment by "competitive exclusion"—they secrete large amounts of acids (acetic, formic, and lactic acids) which make the area unsuitable for pathogens.

Probiotic Bacteria Help Us in Many Ways

Nutritional	Manufacture vitamins in our foods and bodies: B1, B2, B3, B5, B6, B12, vitamins A and K.
Digestive	Digest lactose. Allow some people with lactose intolerance to eat yogurt and cultured dairy products.
	Help regulate peristalsis and regular bowel movements.
	Digest protein to free amino acids.
	Establish good digestion in infants, preventing colic, diaper rash and gas.
Immune	Produce antibiotics and antifungals which prevent colonization of harmful bacteria and fungus.
	Manufacture essential fatty acids, 5–10% of all short-chained fatty acids.
	Increase the number of immune system cells.
	Create lactic acid which balances intestinal pH.
	Break down bacterial toxins and prevent production of bacterial toxins and colitis.
	Have anti-tumor and anti-cancer effects.

Protect from xenobiotics like mercury, pesticides, radiation and harmful pollutants.
Break down bile acids.
Manufacture hydrogen peroxide which has antiseptic effects.

Heart Play a role in normalization of serum cholesterol and triglycerides.

Metabolism Break down and rebuild hormones.
Promote healthy metabolism.
Convert flavinoids (useful as anti-tumor factors and to reduce inflammation) to useable forms.

Benefits of Bifidobacteria[4]

1. Prevent colonization of the intestine by pathogenic bacteria and yeasts by protecting the integrity of the intestinal lining.
2. Produce acids which keep the pH balance in the intestine. This acid environment prevents disease-producing microbes from getting a foothold.
3. Lessen side effects of antibiotic therapy.
4. Are the primary bacteria in infants, which helps them grow.
5. Inhibit growth of bacteria which produce nitrates in the bowel. Nitrates are bowel-toxic and can cause cancer.
6. Help prevent production and absorption of toxins produced by disease-causing bacteria, which reduces the toxic load of the liver.
7. Manufacture B-complex vitamins.
8. Help regulate peristalsis and bowel movements.
9. Prevent and treat antibiotic-induced diarrhea.

Benefits of Lactobacillus acidophillus

1. Prevent overgrowth of disease-causing microbes: candida species, *E. coli*, *H. pylori* and salmonella.
2. Prevent and treat antibiotic-associated diarrhea.
3. Aid digestion of lactose and dairy products.
4. Improve nutrient absorption.

5. Maintain integrity of intestinal tract and protect against macromolecules entering bloodstream and causing antigenic response.
6. Lessen intestinal stress from food poisoning.
7. Acidify intestinal tract; low pH provides a hostile environment for pathogens and yeasts.
8. Help prevent vaginal and urinary tract infections.

In addition to these nutritional and digestive benefits, probiotics enhance immune function. They manufacture antibiotics, like acidophilin produced by acidophillus, which are effective against many types of bacteria, including Streptococcus and staph. *Lactobacillus acidophillus* and *Lactobacillus bulgaricus* have been shown to be effective in laboratory testing against the following pathogens: *Bacillus subtilis, Clostridium botulinum, Clostridium perfringens, Escherichia coli, Proteus mirabilis, Salmonella enteridis, Salmonella typhimurium, Shigella dysenteriae, Shigella paradysenteriae, Staphylococcus aureus, Staphylococcus faecalis.*[5]

Candida albicans, a fungus which causes infections in nails and eyes, thrush and "yeast infections," is controlled by acidophillus. This works in at least two ways. First, acidophillus bacteria ferment glycogen into lactic acid which changes the pH of the intestinal tract. Since Candida and many other disease-causing microbes thrive in alkaline environments, this action discourages many disease-producing microbes. Second, specific strains of Lactobacillus produce hydrogen peroxide which kills Candida directly. Studies show that supplementation with a hydrogen peroxide-producing strain of acidophilus, DDS-1, reduced the incidence of antibiotic-induced vaginal yeast infections threefold. Other probiotics have antitumor and anticancer effects. Probiotics also help us metabolize foreign substances, like mercury and pesticides, and protect us from damaging radiation and harmful pollutants.[6]

Friendly bacteria also help us in other ways. Studies have repeatedly shown that Lactobacillus bacteria can help normalize cholesterol levels. Probiotics also rebuild and break down hormones like estrogen. Probiotics aid digestive function, improve peristalsis and help normalize bowel transit time. Finally, bacterial balance is essential for healthy metabolism. Many super-thin people have been able to gain weight when their bacteria were rebalanced, although the mechanism is not yet understood.

L. bulgaricus and *Streptococcus thermophilus* are two other friendly inhabitants of our digestive tracts. Transient residents of the digestive tract, these flora are not native to it. They "vacation" in us for up to 12 days, which gives them time to have a beneficial effect on the intestinal ecology. Their most obvious function is enhancing the production of bifidobacteria. They have also been shown to have antitumor effects, and *L. bulgaricus* has antibiotic and antiherpes effects as well.[7] They are found in cultured dairy products or can be taken supplementally.

Saccharomyces boulardii is a friendly fungus which enhances levels of secretory IgA. In France it's called "yeast against yeast." Clinically, it has been found to be useful for clearing the skin and controlling diarrhea caused by antibiotics, *Clostridium difficile* and traveling. Saccharomyces has also been used effectively in people with Crohn's disease, significantly reducing the number of bowel movements. Animal studies indicate that it may be used against toxins formed by cholera bacteria.[8]

NOT ALL INTESTINAL BACTERIA ARE FRIENDLY!

Friendly bacteria comprise only a small percentage of our total bacteria. You could call the rest the okay, the bad and the ugly! Most intestinal residents are "commensal" bacteria; they have neither good nor bad effects on how we feel. Other intestinal bacteria are pathogens causing acute illness (Salmonella causes food poisoning) or chronic illness (*Helicobacter pylori* causes ulcers). Disease-causing microbes produce bothersome gas and toxic secretions which irritate the intestinal lining and get absorbed into the bloodstream, making us sick. Some bacteria are extremely virulent and cause sudden and violent illness. Our body's reaction to a strong bacteria like Salmonella is diarrhea, fever, loss of appetite and vomiting. The body is screaming, "Get this stuff out of me!" So it attempts to flush it rapidly out of the body and starve it out. Most disease-causing microbes thrive at human body temperature, while fever kills them by overheating them.

Bacteria that cause chronic illness are generally weak organisms of low virulence. They are often found in small quantities in all of us and have been assumed to be harmless. But when large colonies are given the opportunity to thrive, they can and do cause illness. This type of illness is called intestinal dysbiosis, which is discussed in the next chapter.

Foods and Herbs That Enhance Intestinal Flora

Cultured foods, like yogurt, have significantly increased nutritional content over milk. The bacteria commonly used to produce yogurt—*Lactobacillus bulgaricus* and *L. thermophilus*—produce biotin, a B-complex vitamin. The bacteria make this nutrient for their own benefit, but we benefit as well. By using cultured foods such as sauerkraut rather than cabbage, cottage cheese and yogurt rather than milk, tofu, miso, natto, tamari or shoyu sauce and tempeh rather than soybeans, wine rather than grapes, we get higher levels of vitamins A, B-complex and K.[9] In addition, fermented foods have increased probiotic content, aid digestion and provide health-building enzymes and higher nutritional value.

ENHANCED NUTRIENT CONTENT OF SELECTED DAIRY FOODS

Original Food	Fermented/Cultured Food	Increased Nutrition
Milk	Cheddar cheese	Vitamin B1, 3x
Milk	Cottage cheese	Vitamin B12, 5x
Milk	Yogurt	Vitamin B12, 5-30x
Milk	Yogurt	Vitamin B3, 50x
Skim milk	Low-fat yogurt	Vitamin A, 7-14x

People have long recognized the benefits of fermented foods. Although indigenous people didn't know the science behind their use, they easily noticed the healthful benefits. Sauerkraut, a traditional European food, has a long history of use by people with ulcers and digestive problems. Asian cultures traditionally use pickles and fermented foods as condiments—kimchee, pickled daikon radish or a sweet rice drink called *amasake*. Due to the high incidence of people who are lactose-intolerant around the world, people have relied on cultured dairy products—cottage cheese, kefir, yogurt—for centuries. In India a fermented dairy drink called lassi is a household staple, and in Israel leban, which is similar to yogurt, is served daily. However, people with *Candidiasis* are advised to avoid all fermented foods until they have reestablished their intestinal flora.

Chinese green tea and ginseng also increase friendly flora. Polyphenols are believed to be the enhancing substance in green tea, which has beneficial effects on serum cholesterol, tumors and ulcers. When polyphenols were tested to see their effect on intestinal flora, they increased the number of beneficial bacteria, such as lactobacilli and bifidus, while decreasing the number of disease-causing Clostridium. A significant increase in beneficial flora was also found when ginseng extract was tested in vitro on 107 types of human bacteria.

The traditional Japanese diet takes advantage of several fermented foods—miso, tempeh and tamari or soy sauce—that have antibiotic properties. Miso, a fermented soybean paste, was found to contain 161 strains of aerobic bacteria. Almost all of these were found to compete successfully with *E. coli* and *Staphylococcus aureus*, two main food-poisoning agents. Many lactic acid-producing bacteria were also found. Another positive property of miso is that it helps reduce the negative effects of radioactivity and electromagnetic resonance. Several microbes, including yeast and *L. acidophillus*, are used in the fermentation process of soy sauce, also called shoyu or tamari, which produces a health-giving food. However, in the United States most soy sauce is manufactured from inorganic acids such as hydrochloric acid that break down the soybeans, rather than from living microbes. This type of soy sauce doesn't have the same benefits as traditionally brewed soy sauce.

The Japanese government funds millions of dollars a year in research on the benefits of bifidobacteria and subsidizes companies to put bifidobacteria in their products. Over the past decade Japanese researchers have studied oligosaccharides, the most common of which is fructooligosaccharide (FOS). These sugar molecules rapidly increase the growth of food bifidobacteria. A study done in chickens showed that FOS prevented growth of salmonella. Fructoligosaccharides (FOS) are found in many foods but are especially high in bananas. In fact, British researchers found that banana powder thickened the stomach lining, as opposed to aspirin and Tagamet, a commonly prescribed anti-ulcer medication, which actually thinned the stomach lining.[10] FOS are also found in barley, fruit, garlic, Jerusalem artichoke, onions, soybeans and wheat.

Probiotic Supplements

The composition of intestinal flora usually remains fairly constant in healthy people, but it can become unbalanced by aging, diet, disease, drugs, poor health or stress. In fact, problems due to imbalanced flora have become widespread.

Eating yogurt and cultured dairy products can maintain healthful friendly flora colonies in people who are already healthy, but once disease-producing microbes have colonized, probiotic supplements are necessary.

Taking daily probiotics increases our body's ability to protect itself from illness. Due to their acid-loving nature, they easily survive the high-acid environment of the stomach. Once in the intestinal tract, they colonize and replace less desirable residents. However, not all probiotic supplements provide the same benefits. According to expert Natasha Trenev various strains of acidophillus can differ genetically by as much as 20 percent. This is a huge difference, when you consider that the genetic difference between mice and man is about 2 percent.

What to Look for in a Probiotic Supplement

1. Choose an age-appropriate product. For a baby or toddler, *Bifidobacteria infantis* is appropriate; for children and adults, the most-studied strain of *L. acidophillus* is DDS-1.
2. Because these bacteria are delicate, they must be refrigerated—when shipped, at a store and in your home—to ensure the lifespan of the product and to maintain the greatest potency.
3. Bacteria multiply very quickly, but they need enough food and protection to survive the trip through the stomach and into the intestinal tract. Many manufacturers include FOS or other sugars to feed the probiotics; check to make sure they are included.
4. There is controversy about whether to buy bacteria mixed in one jar or to buy them separately. Most manufacturers produce combination supplements with several types of flora. However, if these flora compete with each other for food, they can limit each other's growth. The only truly cooperative bacteria are the two used in making yogurt: *L. bulgaricus* and *L. thermophilus*.[11] Either purchase acido-

philus and bifidobacteria separately or purchase freeze-dried supplements. Freeze-drying puts the flora into suspended animation, keeping them dormant until placed in water.[12]

Dosage

For preventive measures, take about a billion microbes of each protective species once a day. This is usually about a 1/4 to 1/2 teaspoon or 1 to 3 capsules. For therapeutic purposes, take this amount three times a day. If you take probiotic supplements and have sudden bloating, diarrhea, gas or worsening of symptoms, this is not necessarily a bad sign. As the disease-producing bacteria and fungus are killed, they release chemicals that aggravate symptoms. If this happens, begin again with tiny amounts and build up your dosage slowly to avoid the die-off reaction.

Probiotic supplements are also important for the prevention of traveler's diarrhea. If you plan to travel outside the United States, take a probiotic supplement daily. Studies show that it significantly increases your ability to withstand the new microbes you will be exposed to.[13]

Chapter 5

Dysbiosis: A Good Neighborhood Gone Bad

"Within these regions battles rage; populations rise and fall, affected just as we are by local environmental conditions, industry thrives and constant defense is exercised against interlopers and dangerous aliens who may enter unannounced; colonists roam and settle—some permanently, some only briefly, in general we have in miniature many of terrestrial life's vicissitudes, problems and solutions."

NATASHA TRENEV & LEO CHAITOW, *PROBIOTICS*,
THORSONS PUBLICATION, 1990

EARLY IN THIS century, Dr. Eli Metchnikoff popularized the theory that disease begins in the digestive tract due to an imbalance of intestinal bacteria. He called this state "dysbiosis," which comes from "symbiosis," meaning living together in mutual harmony, and "dys," which means *not!* Dr. Metchnikoff was the first scientist to discover the useful properties of probiotics. He won the Nobel Prize in 1908 for his work on lactobacilli and their role in immunity and was a colleague of Louis Pasteur, succeeding him as the director of the Pasteur Institute in Paris.

Dr. Metchnikoff found that the bacteria in yogurt prevented and reversed bacterial infection. (He named it *Lactobacillus bulgaricus* after the long-lived, yogurt-loving peasants of Bulgaria.) His research proved that lactobacilli could displace many disease-producing organisms and reduce the toxins they generated. He believed these endotoxins—toxins produced from substances inside the body—shortened the lifespan. He advocated use of lactobacillus in the 1940s for ptomaine poisoning, a widely used therapy in Europe.

In recent decades Metchnikoff's work has taken a back seat to modern therapies like antibiotics and immunization programs, which scientists hoped would conquer infectious diseases. For instance, due to an aggressive worldwide immunization program, the World Health Assembly formally declared on May 8, 1980, that smallpox had been eradicated worldwide, which was an enormous triumph for science and humankind. But subsequent efforts at eradicating other diseases have been unsuccessful. While parents in America routinely immunize their children against measles, mumps and polio, parents in poor nations are coping with the loss of half their children by age 10. Worldwide we are finding an increase in new and deadly viruses for which there are no vaccines.

Viruses and bacteria are extremely adaptable. In our efforts to eradicate them we have pushed them to evolve. Long before chemists created antibiotics, yeasts, fungi and rival bacteria were producing antibiotics to ward each other off and establish neighborhoods. They became adept at evading each other's strategies and adapting for survival. Because people have used antibiotics prophylactically and indiscriminately in humans and animals, the bacteria have had a chance to learn from it, undergoing rapid mutations. As they shuffle their components, learning new evolutionary dance steps, superstrains of bacteria have been created which no longer respond to any antibiotic treatment. For instance, our immune systems normally detect bacteria by information coded on the cell walls. Now in response to antibiotics some bacteria have survived by removing their cell walls, so they're able to enter the bloodstream and tissues unopposed, causing damage in organs and tissues.[1] Bacteria can also turn on specific genes when exposed to specific antibiotics.

Resistant strains of bacteria are communicating with each other, and passing resistance information on to other types of bacteria. For example, we now have antibiotic-resistant gonorrhea, leprosy, staph and strep. Similarly, many bacteria that cause disease primarily in the digestive tract—cholera, dysentery, *E. coli*, Enterobacteriaceae, *Enterococcus faecium*, Klebsiella, Proteus, Pseudomonas, Salmonella, *Serratia marcenscens* and Shigella—have mutated to become resistant to specific antibiotics.[2] In 1992, 13,300 hospital patients died of infections that resisted every drug doctors tried.[3]

The Sexual Revolution also helped the evolution of microbes. Multiple sex partners allowed bacteria, fungi and virus more opportunity to replicate, increasing the possibility of mutation. Some of these mutant microbes became better at causing infection or deepening illness. Given people's rapid movement between countries, these new microbes are spread quickly throughout the world, increasing the risk of even more mutation and enhancement of the microbial defense system.

In *The Coming Plague*, Laurie Garrett spends nearly 700 pages discussing microbes, their increasing virulence and the devastation of their epidemics. There is little optimism in her solutions. But she has neglected half the equation: If microbes are becoming more resistant and virulent, we must increase our own resistance and strength to outsmart them. We must boost immune function so that people will be less receptive to infection. We need to take a new look at Metchnikoff's work—at probiotic bacteria and the many immune-strengthening benefits they confer.

Optimum nutrition is a logical starting point. Nutrients,

by definition, are essential for growth, immune function, repair and maintenance of our bodies. But the SAD diet of Americans will not protect our immune function. New research has shown that diets deficient in just one nutrient, in this case either selenium or vitamin E, could cause a benign virus to mutate to a disease-producing organism.[4] Today, less than 80 percent of women and children meet the RDA for vitamin E and zinc. Less than 80 percent of American women meet the RDA for magnesium.[5] A study of high school runners found that 45 percent of girls and 17 percent of boys were deficient in iron, while 31 percent of female college athletes were also found to be iron-deficient. A recent survey of nutritional status in elderly Americans showed that 25 percent of all elderly patients were malnourished, and 50 percent of all hospitalized elderly suffer from malnutrition.[6]

Dysbiosis weakens our ability to protect ourselves from disease-causing microbes, which are generally composed of low-virulence organisms. Unlike Salmonella, which causes immediate food-poisoning reactions, low-virulence microbes are insidious. They cause chronic problems which go undiagnosed in the great majority of cases. If left unrecognized and untreated, they become deep-seated. For instance, if joint pain is the most obvious symptom, we stop movement. If we have diarrhea, then we try to eat differently. In other cases, the effects are much more severe, resulting in chronic fatigue syndrome or colon disease. The dysbiosis causes us to adapt our lifestyle to new limitations.

Because most doctors in our culture do not yet recognize dysbiosis, symptoms are treated with medication, but the underlying cause is never dealt with, and ultimately people do not get well. Published research has listed dysbiosis as the cause of arthritis, autoimmune illness, B-12 deficiency, chronic fatigue, cystic acne, the early stages of colon and breast cancer, eczema, food allergy/sensitivity, inflammatory bowel disease, irritable bowel syndrome, psoriasis and steatorrhea. These problems were previously unrecognized as being microbial in origin. Common dysbiotic bacteria are Aeromonas, Citrobacter, Helicobacter, Klebsiella, Salmonella, Shigella, *Staphylococcus aureus*, Vibrio and Yersinia. Helicobacter, for example, is commonly found in people with ulcers. Citrobacter is implicated in diarrheal diseases. A common dysbiosis culprit, the Candida fungus, causes a wide variety of symptoms that range from gas and bloating to depression, mood swings and PMS (premenstrual syndrome).

SOME BACTERIA AND THEIR RELATIONSHIP TO DIGESTIVE & AUTOIMMUNE PROBLEMS

Bacteria	Gastrointestinal Effect
Citrobacter freundii & diversus	Implicated in diarrheal diseases. Possess an antigen similar to some strains of salmonella which may cause cross-reactive immune reactions. May invade the intestinal mucosa, causing irritation and inflammation.
E. coli	Indicate intestinal flora imbalance when low levels found in stool samples. Look for diet, drugs, environmental stress, excessive yeast, parasites and pathogens as possible causes.
Enterobacter	Associated with diarrhea in children, when found in large numbers. Indicator of imbalanced flora.
Klebsiella	Can cause bacteremia, cystitis, pneumonia, prostatitis. Linked with such autoimmune diseases as ankylosing spondylitis and myasthenia gravis. Usually asymptomatic in the intestine, although it can cause diarrhea. Generally very resistant to antibiotics. Infections often acquired in hospitals, mainly in patients with low resistance.
Lactobacillus	Low levels indicate flora imbalance. Look for diet, drugs, environmental stress, exces-

	sive yeast, parasites and pathogens as possible causes.
Nonlactose E. coli	90% of normal strains of *E. coli* produce lactase, which ferments lactose. Half of the pathogenic strains fail to do so. May be causing symptoms if found in a stool culture in a patient with diarrheal symptoms.
Proteus	Normally found in flora of feces, soil and water. In large numbers, may promote diarrhea and GI distress. *Proteus vulgaris* may be involved in the initiation of myasthenia gravis. Also associated with ankylosing spondylitis and rheumatoid arthritis.
Pseudomonas	May be acquired from contaminated water and lead to diarrhea and GI distress.
Salmonella	Most often reported cause of food poisoning from eggs, milk products and raw meat; infects 38% of commercial uncooked chickens. May occur without symptoms in a carrier state. May also invade the mucosa and present as enteric fever, a focal disease with or without septicemia or gastroenteritis. Increasing health concern.
Staphylococcus aureus	Major cause of food poisoning, with abrupt onset of nausea and vomiting; may include diarrhea and abdominal cramps. Also implicated in pseudomembranous colitis and toxic-shock syndrome.
Yersinia enterocolitica	Implicated as a significant cause of gastroenteritis. Variable symptoms may resemble ulcerative colitis. Antigenic determinants, cross-reactive to the thyroid plasma membrane, may result in Grave's disease. Also associated with such autoimmune diseases as arthritis, erythema nodosum or Sjögren's syndrome.

WHAT CAUSES DYSBIOSIS?

There are many causes of dysbiosis, but we generally bring it on ourselves. Constant high levels of stress, exposure to manufactured chemicals, poor food choices, oral contraceptives, surgery and use of antibiotics and pain killers all change the healthy balance of the digestive tract.

The most common cause of dysbiosis is the use of antibiotics, which change the balance of intestinal microbes. Not terribly specific, antibiotics simultaneously kill both harmful and helpful bacteria throughout our digestive system, mouth, vagina and skin, leaving the territory to bacteria, parasites, viruses and yeasts that are resistant to the antibiotic that was used. In a healthy gut, parasites may be present in small numbers and not cause symptoms, but if allowed to overgrow they can cause diarrhea, illness and weight loss. Most people can recover fairly easily from a single round of antibiotics, but even those with strong constitutions have trouble regaining balance from repeated use of antibiotic drugs.

These microbes produce toxins that cause symptoms. The bacteria form chemicals that are poisonous to the cells around them and to the person they live in. A wide variety of substances are produced, including amines, ammonia, hydrogen sulfide, indoles, phenols, and secondary bile acids.[8] These substances may hurt the intestinal lining directly by damaging the brush border and become absorbed into the bloodstream, causing systemwide effects. Initially, our body rushes white blood cells to the injured tissue to eat up the bacteria and carry away the debris via the lymphatic system. Inflammation, pain and swelling are nature's message to stop and let your body heal. But we often ignore this basic instinct and reach for pain medication so we can continue our lives. If the pain and inflammation were initially caused by microbes and you never dealt with the cause, more endotoxins will be produced, causing chronic pain and inflammation, and setting up a continuing cycle.

Often, the pain medications we take become a factor in

the continuation and severity of the problem. Stephen Barrie, N.D., calls it a Machiavellian theme: The drug causes damage to the intestinal lining, causing more inflammation, irritation and pain, so we take more pain medication, which causes further damage. The most common pain medications fall into two groups: steroidal drugs, like cortisone and prednisone, and nonsteroidal anti-inflammatories (NSAIDs), like aspirin, ibuprofen and indomethacin. Let's look at the long-term effects of these medications on the digestive system and their consequences on health.

Corticosteroids are naturally produced by the adrenal glands. Synthetic corticosteroids like cortisone and prednisone are two of the most effective emergency drugs, used for a multitude of problems, including allergies, arthritis, asthma, Crohn's disease, eczema, lupus, poison ivy, psoriasis and ulcerative colitis. They are generally prescription medications, but weaker over-the-counter remedies can be purchased everywhere. Because cortisone and prednisone have such powerful anti-inflammatory effects, they are used long-term by people with chronic illness. But long-term use of cortisone and prednisone depresses the immune system, causing side effects like lowered resistance to infection and parasites, stomach and duodenal ulcers, thinning of bones and dozens of other problems. Steroids are contraindicated for anyone who has a fungal infection because they provide excellent nourishment for fungi. Yet, many people with Candida infections go undiagnosed. In turn, Candida damages the intestinal lining, causing a wide variety of symptoms, including bloating, chronic fatigue, constipation, depression, diarrhea, fatigue, hypoglycemia and premenstrual syndrome, to name but a few. Because corticosteroids are so strong, they suppress your body's ability to work through an illness on its own. They are best suited for an emergency, not daily use.

NSAIDs like aspirin, ibuprofen and indomethacin work by blocking prostaglandins, which are small protein messengers that circulate throughout the body. Some prostaglandins cause pain and inflammation; others cause healing and repair. NSAID drugs block all prostaglandins. The pain is gone, but the healing process is blocked. Since the intestinal lining repairs and replaces itself every three to five days, prolonged use of NSAIDs blocks its repair. The GI side-effects are well known: the lining becomes weak, inflamed and "leaky, causing leaky gut syndrome, or intestinal permeability. NSAID use also increases the risk of ulcers of the stomach and duodenum. These drugs also cause bleeding, damage to the mucous membranes of the intestines and GI inflammation.

Because aspirin is hard on the stomach lining, doctors and advertisers have advised us to switch to anti-inflammatory drugs like ibuprofen. What we aren't told is NSAIDs cause irritation and inflammation of the intestinal tract, leading to colitis and relapse of ulcerative colitis. NSAIDs can cause bleeding and ulceration of the large intestine and may contribute to complications of diverticular disease.[9] Even with moderate use, NSAIDs increase gut permeability. In fact, use of NSAIDs, steroids, antacids and antibiotics are probably the greatest contributors to leaky gut syndrome.

Poor diet also contributes to dysbiosis. A diet high in fat, sugar and processed foods may not have enough nutrients to optimally nourish the body or repair and maintain the digestive organs. The nutrients most likely to be lacking are the antioxidants—vitamins C and E, beta-carotene, co-enzyme Q10, glutathione, selenium, the sulfur amino acids and zinc—the B-complex vitamins, calcium, essential fatty acids and magnesium.

Poor ileocecal valve function can contribute to dysbiosis. The valve's job is to keep waste matter in the colon from mixing with the useful material that is still being digested and absorbed in the small intestine. When this valve is stuck, either open or closed, dysbiotic problems can occur. Chiropractors can adjust the ileocecal valve to alleviate this problem.

Many other factors contribute to dysbiosis. Low levels of hydrochloric acid (HCl) in the stomach encourage bacterial overgrowth. Poor transit time in the intestinal tract also encourages proliferation of bacteria. For example, in 24 hours one *E. coli* bacterium produces nearly 5,000 identical bacteria! The longer they sit inside us, the greater their potential to colonize.

Patterns of Dysbiosis

There are four commonly recognized patterns of dysbiosis: putrefaction, fermentation, deficiency and sensitization.

Putrefaction Dysbiosis

This is the most common type and occurs when food is not well digested, essentially rotting inside us. We may feel this as bloating, discomfort and indigestion. The typical American high fat, high animal-protein, low fiber diet predisposes people to putrefaction. It causes an increase of bacteroides bacteria, a decrease in beneficial bifidobacteria and an increase in bile production. Bacteroides cause vitamin B-12 deficiency by uncoupling the B-12 from the intrinsic factor necessary for its use. Other bacteria normally make vitamin B-12 directly for their own purposes, but it too becomes unavailable to us. The most common signs of B-12 deficiencies are depression, diarrhea, fatigue, memory loss, numbing of hands and feet, sleep disturbances and weakness. Vitamin B-12 deficiency is commonly seen in older people due to poor hydrochloric acid levels in the stomach.

Poor bifidobacteria levels decrease our body's ability to resist infection, affect bowel health and reflect a decreased production of B-complex vitamins. Research has implicated putrefaction dysbiosis with breast and colon cancer. Bacterial enzymes change bile acids into 33 substances formed in the colon that are tumor promoters. Bacterial enzymes recreate estrogens that were already broken down. These estrogens are reabsorbed into the bloodstream, increasing estrogen levels and estrogen-dependent breast cancers.[10] Putrefaction dysbiosis can be corrected by increasing high fiber foods, fruits, vegetables and grains in our diet while decreasing meats and fats.

Fermentation Dysbiosis

This is characterized by bloating, constipation diarrhea, fatigue and gas. People with fermentation dysbiosis have faulty digestion of carbohydrates: sugars, fruit, beer, wine, grains and fiber. Fermentation of carbohydrates provides food for multiplying bacteria and produces hydrogen and carbon dioxide gases and short-chained fatty acids, which can be tested. Fermentation dysbiosis is often characterized by flora that have been taken over by Candida fungi or other disease-causing microbes. These microbes damage the intestinal brush borders (microvilli) and lead to increased intestinal permeability. People with fermentation dysbiosis need to strictly avoid carbohydrate-containing foods. It is recommended that they receive therapeutic support to help regain bowel health. Up to 2000 mg of citrus seed extract can be used daily. Use of probiotic supplements and repair of the intestinal mucosa are essential.

Deficiency Dysbiosis

This is characterized by a lack of beneficial flora like Bifidobacteria and Lactobacillus. Use of antibiotics and low-fiber diets are the primary cause of poor flora. Deficiency dysbiosis is often seen in people with irritable bowel syndrome and food sensitivities. Deficiency dysbiosis is often coupled with putrefaction dysbiosis, and the treatment is the same. To restore balance, supplementation with beneficial flora and fructooligosaccharides (FOS) is also recommended. Up to 2000 mg of citrus seed extract may be used daily.

Sensitization Dysbiosis

This occurs when the immune system reacts with abnormal or aggravated responses to the digestive process. Microbes in the gut and foods produce exotoxins that irritate the gut lining. Our bodies recognize these toxins as foreign substances and produce antibodies which signal the immune system to get rid of them. Unfortunately, this local reaction may be the cause of some autoimmune diseases. Rheumatoid arthritis, ankylosing spondylitis and perhaps skin diseases like eczema and psoriasis are often the result.

The process works like this: Our body initially reacts to a "bug" of some sort, but the antibodies formed to fight it are the perfect keys that fit into *gene markers*. These markers make us more susceptible to certain diseases. The mechanism by which these are turned on or off is unclear, but it appears that specific microbes cause disease *only* in people with specific gene markers who are exposed to specific microbes, which are harmless in everyone else. Our body senses that as a systemic problem rather than just some imbalanced microbes in our intestines. It reacts with an all-out antibody reaction, attacking the body itself—an autoimmune reaction. What originated as a local infection becomes an autoimmune illness. For instance, rheumatoid arthritis has been linked to a prevalence of a bacteria called Proteus which mimics gene marker HLADR4; 50 to 75 percent of people with rheumatoid arthritis have this gene marker.

People with sensitization dysbiosis often have food intolerances, leaky gut syndrome and increasing sensitivity to foods and the environment. Symptoms may include acne, bowel or skin problems, connective tissue disease and psoriasis. Sensitization dysbiosis may accompany fermentation dysbiosis, and similar treatments may be helpful. Replacing probiotic bacteria and repair of the intestinal mucosa are essential.

CANDIDA: THE MASQUERADER

The most prevalent and obvious form of dysbiosis is candidiasis, a fungal infection. Candida is found in nearly everyone and in small amounts is compatible with good health. Candida is usually controlled by friendly flora, our immune defense system and intestinal pH. When the bacteria that normally balance candida are killed, it causes what's commonly known as a yeast infection because it produces a smell like yeasty bread. Also called a yeast infection in the vagina and a fungal infection in the nailbeds and the eyes, it's called thrush in the throat. *Candida albicans* is the usual offender, but other species of candida fungus may cause health problems as well.

In the early 1980s, Orion Truss, M.D. noticed that many of his patients' other problems resolved when he treated them with nystatin for fungal problems. Indeed, one patient had a complete reversal of multiple sclerosis. After hearing Dr. Truss lecture on his findings, Abraham Hoffer, a doctor in the field of orthomolecular psychiatry, tried his first yeast protocol on a psychiatric patient who had suffered from depression for many years. One month after initiating Truss's program she was mentally and emotionally normal.[11]

What is less well recognized in contemporary medical practice is that candida can also colonize in the digestive tract, causing havoc everywhere in the body. Candida colonies produce powerful toxins that are absorbed into the bloodstream and affect our immune system, hormone balance and thought processes. The most common symptoms are abdominal bloating, anxiety, constipation, diarrhea or both, depression, environmental sensitivities, fatigue, feeling worse on damp or muggy days or in moldy places, food sensitivities, fuzzy thinking, insomnia, low blood sugar, mood swings, premenstrual syndrome, recurring vaginal or bladder infections, ringing in the ears, and sensitivities to perfume, cigarettes or fabric odors. Although these symptoms are the most prevalent, candida has many faces and many types of symptoms can occur.

American physicians recognize candida in many forms, but in general they have been unwilling to recognize it in a systemic form. For those who do, it is a common diagnosis. In my own experience, I have found candida to be an underlying cause in many clients. Abraham Hoffer, M.D. states that one third of the world's population is affected by candidiasis.

Candida infections are usually triggered by use of antibiotics, birth control pills or steroid medications. These drugs change the balance of the intestinal tract, kill the bacteria that keep candida in check, and the fungus quickly takes hold. Candida are like bullies that push their way into the intestinal lining, destroying cells and brush borders. Greater numbers of candida produce greater amounts of toxins, which further irritate and break down the intestinal lining. This damage allows macromolecules of partially digested food to pass through. The macromolecules are the perfect size for antibodies to respond to. Your immune system then goes on alert for these specific foods so the next time you eat them, your antibodies will be waiting! The net result is increased sensitivity to foods and other food substances and the environment.

There are many, many therapeutic programs that people may follow to rid themselves of candida. The standard therapeutic diet allows beef, chicken, fish, eggs, poultry, yogurt, vegetables, oils, nuts and seeds. No alcoholic beverages, fruit or dried fruits, flours, grains, mushrooms, sugar, vinegar or yeasted breads are allowed. When I had candida, I found that the worst foods for me were beer, sugar and dried fruit. A sip of beer triggered symptoms within 20 minutes. At that time I was a banana-holic, consuming three to five daily. I now know that bananas are loaded with fructooligosaccharides, which feed lactobacillus and bifidobacteria but not candida! Due to biochemical individuality, other people find other foods and substances to be the worst triggers for them. People are advised to go on a strict diet for three weeks and then reintroduce foods one at a time to test for reactions. If you know you're sensitive to a specific food, avoid it.

Candida often causes bowel problems—either diarrhea or constipation or both alternately—so it is essential that bowel transit time be normalized. Adding psyllium seed husks (found in fiber supplements) and probiotic supplements greatly helps to normalize this condition. Russel Jaffe, M.D. uses vitamin C flushes, which are described in Chapter 9, Detoxification.

Many substances are helpful in killing off candida. Garlic is my personal favorite—use lots of it raw as a vaginal suppository (make sure not to nick the garlic or it can sting!) or take in capsule form. Capryllic acid from coconuts, oleic acid from olive oil, pau d'arco and grapefruit seed extract are all valuable agents for killing candida. Mathake, a South American herb, has been found to be extremely effective. While it isn't necessary to use all these health enhancers, you can buy many of them in combination products in health food stores or from health professionals.

When the candida are killed, the protein fragments and endotoxins released trigger an antibody response. This can initially produce a worsening of the person's symptoms and is commonly known as a die-off reaction, or a Herxheimer reaction. Therefore, it is important to begin therapeutics gently with small doses and gradually increase. If your symptoms are still initially aggravated, cut back and gradually increase supplements. Most people begin to feel dramatically better within two weeks. If you don't, you're probably not dealing with a candida problem. Ask your health professional to make therapeutic recommendations.

I recently advised a client to take probiotic supplements for peeling skin and burning sensation on her feet. Although she scored rather low on the candida questionnaire, her symptoms fit those of candida. I advised her to take probiotics slowly, beginning with 1/4 teaspoon daily and working up gradually to a full teaspoon once or twice a day. Six days later, she told me her feet were improving, but she had a horrible headache every time she took a teaspoon of probiotics. After she began more slowly, she was soon relieved of both problems.

The following Candida Questionnaire can help you determine if candida is a factor in your own health. The bibliography lists several informative books on candida.

Exercise 4

YEAST QUESTIONAIRE—ADULT[12]

Answering these questions and adding up the scores will help you decide if yeasts contribute to your health problems. Yet you will not obtain an automatic "yes" or "no" answer.

For each "yes" answer in Section A, circle the point score in that section. Total your score and record it at the end of the section. Then move on to sections B and C and score as indicated.

Add the total of your scores to get your *Grand Total Score.*

SECTION A: HISTORY

	Point score
1. Have you taken tetracyclines (Sumycin®, Panmycin®, Vibramycin®, Minocin®, etc.) or other antibiotics for acne for 1 month (or longer)?	35
2. Have you, at any time in your life, taken other "broad spectrum" antibiotics* for respiratory, urinary or other infections (for 2 months or longer, or in shorter courses 4 or more times in a 1-year period?)	35
3. Have you taken a broad spectrum antibiotic drug* even a single course?	6
4. Have you, at any time in your life, been bothered by persistent prostatis, vaginitis or other problems affecting your reproductive organs?	25

*Including Keflex®, ampicillin, amoxicillin, Ceclor®, Bactrim® and Septra®. Such antibiotics kill off "good germs" while they're killing off those which cause infection.

5. Have you been pregnant . . .	
2 or more times?	5
1 time?	3
6. Have you taken birth control pills . . .	
For more than 2 years?	15
For 6 months to 2 years?	8
7. Have you taken prednisone, Decadron® or other cortisone-type drugs . . .	
For more than 2 weeks?	15
For 2 weeks or less?	6
8. Does exposure to perfumes, insecticides, fabric shop odors and other chemicals provoke . . .	
Moderate to severe symptoms?	20
Mild symptoms?	5
9. Are your symptoms worse on damp, muggy days or in moldy places?	20
10. Have you had athlete's foot, ring worm, "jock itch" or other chronic fungus infections of the skin or nails? Have such infections been . . .	
Severe or persistent?	20
Mild to moderate?	10
11. Do you crave sugar?	10
12. Do you crave breads?	10
13. Do you crave alcoholic beverages?	10
14. Does tobacco smoke *really* bother you?	10
Total Score, Section A......................................	

SECTION B:
MAJOR SYMPTOMS: ______________________

For each of your symptoms, enter the appropriate figure in the Point Score column:

If a symptom is *occasional or mild*............................score 3 points
If a symptom is *frequent and/or moderately severe*...score 6 points
If a symptom is *severe and/or disabling*...................score 9 points

Add total score and record it at the end of this section.

	Point score
1. Fatigue or lethargy	
2. Feeling of being "drained"	
3. Poor memory	
4. Feeling "spacey" or "unreal"	
5. Depression	
6. Inability to make decisions	
7. Numbness, burning or tingling	
8. Muscle aches or weakness	
9. Pain and/or swelling in joints	
10. Abdominal pain	
11. Constipation	
12. Diarrhea	
13. Bloating, belching or intestinal gas	
14 Troublesome vaginal burning, itching or discharge	
15. Persistent vaginal burning or itching	
16. Prostatitis	
17. Impotence	
18. Loss of sexual desire or feeling	
19. Endometriosis or infertility	
20. Cramps and/or other menstrual irregularities	
21. Premenstrual tension	
22. Attacks of anxiety or crying	
23. Cold hands or feet and/or chilliness	
24. Shaking or irritable when hungry	
Total Score, Section B..................................	

SECTION C:
OTHER SYMPTOMS:* ______________________

For each of your symptoms, enter the appropriate figure in the point score column:

If a symptom is *occasional or mild*............................ score 1 point
If a symptom is *frequent and/or moderately severe*... score 2 points
If a symptom is *severe and/or disabling*.................... score 3 points
Add total score and record it at the end of this section.

*While the symptoms in this section commonly occur in people with yeast-connected illness they are also found in other individuals.

	Point score
1. Drowsiness	
2. Irritability or jitteriness	
3. Incoordination	
4. Inability to concentrate	
5. Frequent mood swings	
6. Headache	
7. Dizziness/loss of balance	
8. Pressure above ears . . . feeling of head swelling	
9. Tendency to bruise easily	
10. Chronic rashes or itching	
11. Numbness, tingling	
12. Indigestion or heartburn	
13. Food sensitivity or intolerance	
14. Mucus in stools	
15. Rectal itching	
16. Dry mouth or throat	
17. Rash or blisters in mouth	
18. Bad breath	
19. Foot, body or hair odor not relieved by washing	
20. Nasal congestion or postnasal drip	
21. Nasal itching	
22. Sore throat	
23. Laryngitis, loss of voice	
24. Cough or recurrent bronchitis	
25. Pain or tightness in chest	
26. Wheezing or shortness of breath	
27. Urgency or urinary frequency	
28. Burning on urination	
29. Spots in front of eyes or erratic vision	
30. Burning or tearing of eyes	
31. Recurrent infections or fluid in ears	
32. Ear pain or deafness	
Total Score, Section C	
Total Score, Section A..................................	
Total Score, Section B..................................	
GRAND TOTAL SCORE	

The Grand Total Score will help you and your physician decide if your health problems are yeast connected. Scores in women will run higher as 7 items in the questionnaire apply exclusively to women, while only 2 apply exclusively to men.

Yeast-connected health problems are almost certainly present in women with scores *over 180* and in men with scores *over 140.*

Yeast-connected health problems are probably present in women with scores *over 120* and in men with scores *over 90.*

Yeast-connected health problems are possibly present in women with scores *over 60* and in men with scores *over 40.*

With scores of less than 60 in women and 40 in men, yeasts are les apt to cause health problems.

YEAST QUESTIONAIRE—CHILD

Circle appropriate point score for questions you answer "yes." Total your score and record it at the end of the questionnaire.

	Point score
1. During the two years before your child was born, were you bothered by recurrent vaginitis, menstrual irregularities, premenstrual tension, fatigue, headache, depression, digestive disorders or "feeling bad all over"?	30
2. Was your child bothered by thrush? (Score 10 if mild, score 20 if severe or persistent)	10 20
3. Was your child bothered by frequent diaper rashes in infancy? (Score 10 if mild, 20 if severe or persistent)	10 20
4. During infancy, was your child bothered by colic and irritability lasting over 3 months? (Score 10 if mild, 20 if moderate or severe)	10 20
5. Are his symptoms worse on damp days or in damp or moldy places?	20
6. Has your child been bothered by recurrent or persistent "athlete's foot" or chronic fungus infections of his skin or nails?	30

Question	Score
7. Has your child been bothered by recurrent hives, eczema or other skin problems?	10
8. Has your child received:	
a. 4 or more courses of antibiotic drugs during the past year? Or has he received continuous "prophylactic" courses of antibiotic drugs?	80
b. 8 or more courses of "broad-spectrum" antibiotics . . . (such as amoxicillin, Keflex®, Septra®, Bactrim®, or Ceclor®) during the past three years?	50
9. Has your child experienced recurrent ear problems?	10
10. Has your child had tubes inserted in his ears?	10
11. Has your child been labeled "hyperactive"? (Score 10 if mild, 20 if moderate or severe)	10 20
12. Is your child bothered by learning problems (even though his early developmental history was normal)?	10
13. Does your child have a short attention span?	10
14. Is your child persistently irritable, unhappy and hard to please?	10
15. Has your child been bothered by persistent ot recurrent digestive problems, including constipation, diarrhea, bloating or excessive gas? (Score 10 if mild; 20 if moderate; 30 if severe)	10 20 30
16. Has he been bothered by persistent nasal congestion, cough and/or wheezing?	10
17. Is your child unusually tired or unhappy or depressed? (Score 10 if mild, 20 if severe)	10 20
18. Has your child been bothered by recurrent headaches, abdominal pain, or muscle aches? (Score 10 if mild, 20 if severe)	10 20
19. Does your child crave sweets?	10
20. Does exposure to perfume, insecticides, gas or other chemicals provoke moderate to severe symptoms?	30
21. Does tobacco smoke really bother him?	20
22. Do you feel that your child isn't well, yet diagnostic tests and studies haven't revealed the cause?	10
TOTAL SCORE	

Yeasts possibly play a role in causing health problems in children with scores of 60 or more.

Yeasts probably play a role in causing health problems in children with scores of 100 or more.

Yeasts almost certainly play a role in causing health problems in children with scores of 140 or more.

Chapter 6

LEAKY GUT SYNDROME: THE SYSTEMIC CONSEQUENCES OF FAULTY DIGESTION

"The gut is a major, potential portal of entry into the body for foreign antigens. Only its intact mucosal barrier protects the body from foreign antigen entry and systemic exposure."

RUSSELL JAFFE, M.D., *ELISA/ACT CININICAL UPDATE* 2, NO. 1 (JANUARY 1992)

A HEALTHY INTESTINAL LINING allows only properly digested fats, proteins and starches to pass through so they can be assimilated. At the same time it also provides a barrier to keep out bacterial products, foreign substances and large undigested molecules. This is called the "barrier function" of the gastrointestinal mucosal lining. This surface is often called the "brush border," because under a microscope its villi and microvilli look like bristles on a brush.

The intestinal lining lets substances move across this barrier in several ways. The process of "diffusion" is a simple one: it equalizes the concentrations inside and outside the cells. Diffusion is the way ions of chloride, magnesium, potassium, sodium and free fatty acids pass into the cells. Most nutrients are moved through the brush borders via a process called "active transport." Carrier molecules, all of low molecular weight, transport nutrients like molecular taxis. Amino acids, fatty acids, glucose, minerals and vitamins cross cell membranes through active transport.

In-between cells are junctions called "desmosomes." Normally, desmosomes form tight junctions and do not permit large molecules to pass through. But when the area is irritated and inflamed, these junctions loosen up, allowing larger molecules to pass through. The substances that pass through the intracellular junctions are seen by our immune system as foreign, stimulating an antibody reaction. When the intestinal lining is damaged, larger substances of particle size are allowed to pass directly, again triggering an antibody reaction.

When the intestinal lining is damaged even more, substances larger than particle size—disease-causing bacteria, potentially

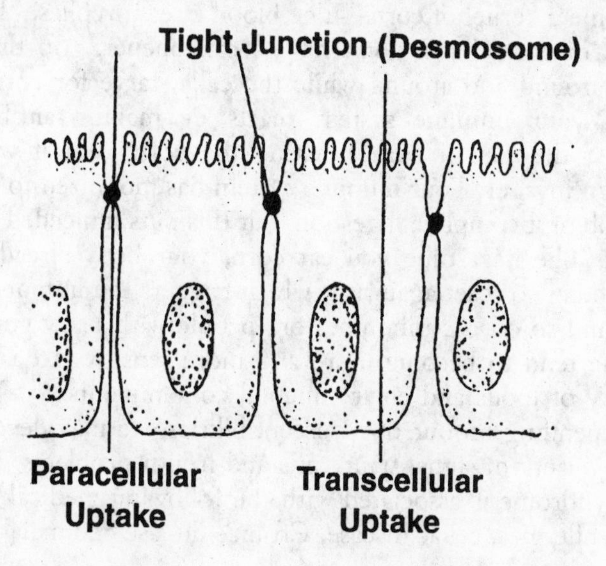

toxic molecules and undigested food particles—are allowed to pass directly through the weakened cell membranes. They go directly into the bloodstream, activating antibodies and alarm substances called cytokines. The cytokines alert our lymphocytes—white blood cells—to battle the particles. Oxidants are produced in the battle, causing irritation and inflammation far from the digestive system. That is the basis for a condition called increased intestinal permeability or leaky gut syndrome.

Intestinal mucus normally blocks bacteria from moving to other parts of the body. But when the cells are leaking, bacteria passes into the bloodstream and throughout the body. When intestinal bacteria colonize in other parts of the body, we call it bacterial translocation, and it is often found in people with leaky gut syndrome. For example, *Blastocystis hominis*, a bacteria that causes GI problems, has been found in the synovial fluid in the knee of an arthritis patient. Surgery or tube feeding in hospitals can also cause bacterial translocation.

Here's how leaky gut syndrome works. Imagine that your cells need a kernel of corn. They are screaming out, "Hey, send me a kernel of corn." The blood stream replies, "I have a can of corn, but I don't have a can opener." So the can goes around and around while the cells starve for corn. Finally, your immune system reacts by making antibodies against the can of corn, treating the corn as if it were a foreign invader. Your immune system has mobilized to finish the job of incomplete digestion, but this puts unneeded stress on it. The next time you eat corn, your body already has antibodies to react against it, which triggers the immune system and so on. As time goes on, people with leaky gut syndrome tend to become more and more sensitive to a wider variety of foods and environmental contaminants.

Depending on our own susceptibilities, we may develop a wide variety of signs, symptoms and health problems. Leaky gut syndrome is associated with the following medical problems: allergies, celiac disease, Crohn's disease and malabsorption syndromes. It is also linked to autoimmune diseases like AIDS, ankylosing spondylitis, asthma, atopy, bronchitis, eczema, food and environmental sensitivities, other allergic disorders, psoriasis, Reiter's syndrome, rheumatoid arthritis, Sjögren's syndrome and skin irritations.[1]

Common Clinical Conditions Associated with Intestinal Permeability[2]

Alcoholism	Aging
Autism	Childhood hyperactivity
Celiac disease	Malnutrition
Chemotherapy	Celiac disease
Crohn's disease	Giardia
Environmental illness	Multiple chemical sensitivities
HIV-positive	Ankylosing spondylitis
Hives	Chronic fatigue syndrome
Acne	Eczema
Cystic fibrosis	Psoriasis
Inflammatory joint disease/ arthritis	Food allergies and food sensitivities
Intestinal infections	Endotoxemia
Malnutrition	
Pancreatic insufficiency	Liver dysfunction
Schizophrenia	Trauma
Thermal injury	NSAIDs enteropathy
Ulcerative colitis	Irritable bowel syndrome

The conditions in the following chart can arise from a variety of causes, but leaky gut syndrome may underlie more classic diagnoses. If you have any of the common symptoms or disorders associated with leaky gut syndrome, ask your physician to order an intestinal permeability test to see if it is causing your problem. This test is best done in conjunction with a comprehensive digestive and stool analysis with parasitology, which checks for digestive function and dysbiosis.

In addition to clinical conditions, people with leaky gut syndrome display a wide variety of symptoms.

SYMPTOMS ASSOCIATED WITH LEAKY GUT SYNDROME[3]

- Abdominal pain
- Asthma
- Chronic joint pain
- Chronic muscle pain
- Confusion
- Fuzzy thinking
- Gas
- Indigestion
- Mood swings
- Nervousness
- Poor exercise tolerance
- Poor immunity
- Recurrent vaginal infections
- Skin rashes
- Diarrhea
- Bed wetting
- Recurrent bladder infections
- Fevers of unknown origin
- Poor memory
- Shortness of breath
- Constipation
- Bloating
- Aggressive behavior
- Anxiety
- Primary biliary cirrhosis
- Fatigue and malaise
- Toxic feelings

Leaky gut syndrome puts an extra burden on the liver. All foods pass directly from the bloodstream through the liver for filtration. The liver "humanizes" the food and either lets it pass or changes it, breaking down or storing all toxic or foreign substances. Water-soluble toxins are easily excreted, but the breakdown of fat-soluble toxins is a two-stage process that requires more energy. When the liver is bombarded by inflammatory irritants from incomplete digestion, it has less energy to neutralize chemical substances. When overwhelmed, it stores these toxins in fat cells, much the same way that we put boxes in the garage or basement to deal with at a later date. If the liver has time later, it can deal with the stored toxins, but most commonly it is busy dealing with what is newly coming in and never catches up. These toxins provide a continued source of inflammation to the body.

What Causes Leaky Gut Syndrome?

There is no single cause of leaky gut syndrome, but some of the most common are chronic stress, dysbiosis, environmental contaminants, gastrointestinal disease, immune overload, overuse of alcoholic beverages, poor food choices, presence of pathogenic bacteria, parasites and yeasts, and prolonged use of NSAIDs.[4] Let's discuss some of these one at a time.

Chronic stress: Prolonged stress changes the immune system's ability to respond quickly and affects our ability to heal. It's like the story of "The Little Boy Who Cried Wolf." If we keep hollering that there's a wolf every time we're late for an appointment or we need to finish a project by a deadline, our bodies can't tell the difference between this type of stress and real stress—like meeting a vicious dog in the woods or a death in the family. Our body reacts to these stressors by producing less secretory IgA (one of the first lines of immune defense) and less DHEA (an anti-aging, antistress adrenal hormone) and by slowing down digestion and peristalsis, reducing blood flow to digestive organs, and productng toxic metabolites. Meditation, guided imagery, relaxation and a good sense of humor can help us deal with daily stresses. We can learn to let small problems and traumas wash over us, not taking them too seriously.

Dysbiosis: The presence of dysbiosis contributes to leaky gut syndrome. Candida push their way into the lining of the intestinal wall and break down the brush borders. They must be evaluated when leaky gut syndrome is suspected. *Blastocystisis hominis*, Giardia, Helicobacter, Salmonella, Shigella, *Yersinia enterocolitica*, amoebas, and other parasites also irritate the intestinal lining and cause gastrointestinal symptoms. People who have or have had digestive illness or liver problems have an increased tendency to leaky gut syndrome. Which came first: the chicken or the egg?

Environmental contaminants: Daily exposure to hundreds of household and environmental chemicals puts stress on our immune defenses and the body's ability to repair. This leads to chronic delay of necessary routine repairs. Our immune systems can only pay attention to so many places at one time. Parts of the body far away from the digestive system are affected. Connective tissue begins to break down, and we lose trace minerals like calcium, potassium and magnesium. Environmental chemicals deplete our reserves of buffering minerals, causing acidosis in the cells and tissue and cell swelling. This is known as leaky cells—like having major internal plumbing problems!

Overconsumption of alcoholic beverages: Alcoholic drinks contain few nutrients but take many nutrients to metabolize. The most noteworthy of these are the B-complex vitamins. In fact, alcoholic beverages contain substances that are toxic to our cells. When alcohol is metabolized in the liver, the toxins are either broken down or stored by the body. Alcohol abuse puts a strain on the liver which affects digestive competency and also damages the intestinal tract.

Poor food choices: Poor food choices contribute to an imbalance of probiotics and pH. An intestinal tract that is too alkaline promotes dysbiosis. Low-fiber diets cause an increase in transit time, allowing toxic byproducts of digestion to concentrate and irritate the gut mucosa. Diets of highly processed foods injure our intestinal lining. Processed foods invariably are low in nutrients and fiber, with high levels of food additives, restructured fats and sugar. These foods promote inflammation of the GI tract. In fact, even foods we normally think of as healthful can be irritating to the gut lining. Milk, an American staple, can be highly irritating to people with lactose intolerance.

Use of medication: As was discussed more fully in Chapter 5, NSAIDs damage brush borders, allowing microbes, partially digested food particles and toxins to enter the bloodstream.

Birth control pills and steroid drugs also create conditions that help feed fungi, which damage the lining. Chemotherapy drugs and radiation therapy can also significantly disrupt GI balance.

Food and Environmental Sensitivities

Food and environmental sensitivities are the result of leaky gut syndrome. The prevalence of these sensitivities is more widely recognized today than in the past; 24 percent of American adults claim they have food and environmental sensitivities.[5] These sensitivities, also called delayed hypersensitivity reactions, differ from true food allergies, also called Type 1 or immediate hypersensitivity reactions.

True food allergies are rare occurrences. They affect 0.3 to 7.5 percent of children and 1 to 2 percent of adults. The foods that most often trigger these reactions are eggs, cows' milk, nuts, shellfish, soy, wheat and white fish.[6] Food allergies trigger reaction of type IgE antibodies which bind to the offending food antigens. The IgE reaction causes the release of cytokines and histamines that results in closing of the throat, hives, itching, respiratory distress, runny nose, skin rashes and sometimes severe reactions of asthma and anaphylactic shock. These symptoms occur within minutes after the food is eaten. Physicians diagnose food allergies through the use of patch skin tests and RAST blood testing. These tests are great for testing for food allergies but do not accurately determine food sensitivities.

Sensitivity reactions, also called "delayed" or "hidden" hypersensitivities, occur when IgA, IgG and IgM antibodies are triggered in response to foods, chemicals and bacterial toxins. The sensitivities cause symptoms which are delayed, taking several hours to several days to appear. This makes tracking them down very difficult. Food and environmental sensitivities cause a wide number of symptoms typical of a leaky gut reaction. Food particles enter the bloodstream through damaged mucosal membranes, the body recognizes them as foreign substances (antigens) and triggers an immune reaction. Prolonged antibody response can overwhelm the liver's ability to eliminate these food antigens. Subsequently, the antigens enter the bloodstream and trigger delayed hypersensitivity response, inflammation, cell damage and disease. Almost any food can cause a reaction, although the most common are beef, citrus, dairy products, egg, pork and wheat. These foods provoke 80 percent of food sensitivity reactions.

SYMPTOMS OF FOOD & ENVIRONMENTAL SENSITIVITY

The following symptoms occur because of many health conditions. Professional evaluation is necessary to uncover the source of these symptoms and to establish if food sensitivities are involved.[7]

HEAD: Chronic headaches, migraines, difficulty sleeping, dizziness

MOUTH & THROAT: Coughing, sore throat, hoarseness, swelling/pain, gagging, frequently clearing throat, sores on gums, lips and tongue

EYES, EARS, NOSE: Runny or stuffy nose, postnasal drip, ringing in the ears, blurred vision, sinus problems, watery and itchy eyes, ear infections, hearing loss, sneezing attacks, hay fever, excessive mucus formation, dark circles under eyes, swollen, red or sticky eyelids

HEART & LUNGS: Irregular heartbeat (palpitations, arrhythmia), asthma, rapid heartbeat, chest pain & congestion, bronchitis, shortness of breath, difficulty breathing

GASTROINTESTINAL: Nausea and vomiting, constipation, diarrhea, irritable bowel syndrome, indigestion, bloating, passing gas, stomach pain, cramping, heartburn

SKIN: Hives, skin rashes, psoriasis, eczema, dry skin, excessive sweating, acne, hair loss, irritation around eyes

MUSCLES & JOINTS: General weakness, muscle/joint aches and pains, arthritis, swelling, stiffness

ENERGY & ACTIVITY: Fatigue, depression, mental dullness and memory lapses, difficulty getting your work done, apathy, hyperactivity, restlessness

EMOTIONS & MIND: Mood swings, anxiety and tension, fear, nervousness, anger, irritability, aggressive behavior, "binge" eating or drinking, food cravings, depression, confusion, poor comprehension, poor concentration, difficulty learning

OVERALL: Overweight, underweight, fluid retention, dizziness, insomnia, genital itch, frequent urination

Additional Signs of Food Sensitivities in Children

In addition to the symptoms listed above, children with food sensitivities may have:

Attention Deficit Disorder
Behavior problems
Learning problems
Reoccurring ear infections

These problems are often not recognized as being related to food sensitivities. Children with these problems will benefit from a food evaluation and environmental sensitivity testing.

Blood testing for IgG4 antibody reactions can help determine sensitivities to a variety of foods and environmental substances. You may want to screen for food allergies with IgE testing at the same time. Some labs do a test for another indicator of delayed hypersensitivity: white blood cell blastogenesis, where lymphocytes are stimulated and produce protein, DNA, and RNA at a rapid rate. Environmental screening panels measure antibody reaction to chemicals commonly found in our homes, yards, workplaces and public places. People often test positive to household cleaning supplies and petroleum-based chemicals. Several laboratories perform antibody testing for foods, dusts, environmental chemicals, heavy metals, molds and pollens. These labs are listed in the Resources section.

If you wish to determine food and chemical sensitivities on your own, you can use the elimination provocation challenge. Only eat foods that you are unlikely to be sensitive to for a week and then add back the foods you normally eat to "challenge" your system. Removal of offending foods calms down symptoms, while careful addition of only one food each two days makes it easier to determine which foods caused the reaction. Although the elimination provocation challenge sounds simple, it can be tricky. People usually have no problem with the elimination part—a restricted food plan for a week is easy. Slowly adding foods back into your diet may be more difficult, because recipes and restaurant foods have many ingredients. Sometimes it's hard to determine which ingredient caused the distress. Reactions that are delayed for a day or two also complicate the situation. It then becomes necessary to remove all suspected foods for four days, and try them again one at a time. If you have the same reaction each time you add the food, you've found the culprit. Unfortunately, if you have sensitivities to one food, you are often sensitive to all foods in the same family. For example, some people who are sensitive to wheat are sensitive to all grains in the grass family. It is common to be sensitive to more than one food or food family.

To cure food and environmental sensitivities, you'll do best with a holistic approach. Begin by avoiding substances you are sensitive to for a period of six months, and your body will gradually stop reacting to most of them. That will help detoxify the body, especially the liver (detoxification programs are discussed in Chapter 9). A comprehensive program of nutritional supplements will help in the healing process. *The Elisa/Act Handbook* reads: "Persons suffering from immune system dysfunction and overload due to delayed hypersensitivity reactions often have a need for even greater supplementation because of poor functioning of the body's normal biochemical pathways."[8] Natural foods, organically grown and nutrient rich, also help repair the body. Exercise programs and use of stress management tools also play a part in recovery. With a holistic program you will find that over time you will become less and less sensitive to foods and the environment.

Part 3

A New Way of Identifying the Problem

Chapter 7

FUNCTIONAL MEDICINE/ FUNCTIONAL TESTING

"The best measure of quality is not how well or how frequently a medical service is given, but how closely the result approaches the fundamental objective of prolonging life, relieving distress, restoring function and preventing disability."

LEMBEKE QUOTED IN JEFFERY BLAND,
FUNCTIONAL MEDICINE JOURNAL, 1993

MEDICAL DISCIPLINES AND medical testing are like the ancient East Indian folktale of the six blind men touching the elephant. We test for what we "see." But each of us is "blind" because no matter how much we know, there is still so much that is unknown to us. Contemporary medicine offers a great deal, but there are other healing disciplines that lead to different explanations and interpretations. By pooling our information, we can have a fuller understanding that leads to better healing outcomes.

There is a new paradigm in nutritional healing called

functional medicine. Its primary concern is finding health problems before they become illnesses. Questionnaires and lab testing are used to determine how well organs and systems are functioning. Functional medicine recognizes that each of us has specific biochemical needs which are determined by our lifestyle, genetic structure and environment. It focuses on restoration of health and health status, rather than progression of disease. It looks for improvement in the quality of life and an increase in a person's healthspan. The ultimate goal is for people to live a healthful existence throughout life.

Modern medicine equates the absence of disease with health. Often people walk into a doctor's office feeling tired and run down, have an exam, a battery of tests, and come back a week later to hear, "Nothing's wrong with you. You're perfectly healthy." While relieved, they still don't feel well. This is where functional, complementary and alternative medicines play a critical role. People often seek "alternative" treatment as a last resort, but chiropractors, naturopathic physicians, nutritionally oriented M.D.s, nutritionists, and other holistic health practitioners offer a different view of health. Their work focuses on wellness and improved function.

As a nutritionist, I measure improved function by how people feel. When a client comes to me with symptoms or a diagnosed medical problem, I listen carefully to what the client tells me, record the information and ask lots of questions. I was taught that if I listened carefully my clients would tell me 90 percent of what I need to know. While I can't do that with each of you, I have included several questionnaires which can help you discover more about yourself, your lifestyle and your body. These questionnaires also provide valuable information for your health care provider.

When I still don't feel I have all the information I need, laboratory testing is my next resource. Tests give accurate measurements that can be compared with what is normal, average and optimal. Each lab test looks for specific patterns and biochemical markers. Depending on their point of view and training, health care providers may order widely differing lab tests. Like blind men looking at an elephant, they don't always see the same thing.

Many years before most illnesses occur, people have a steady, cumulative, measureable decline in body function. To quantify this, holistic health practitioners require different tools than those most commonly used over the past 50 years. By the time you need a cardiogram, you already have heart symptoms. By the time you have a lower or upper GI test, you are experiencing some digestive problems. Even so-called "preventive medical testing" like mammograms and cholesterol screenings are really tests for early detection of disease rather than true prevention. Innovative people in medical laboratories worldwide have been designing tests to determine function rather than disease. The leading laboratories are listed in the Resources.

There are numerous tests for functional medicine. The following are the ones I've found to be most useful for digestive problems. Most of these are laboratory tests, but a few are home tests that you can perform on your own!

Comprehensive Digestive and Stool Analysis

Comprehensive digestive and stool analysis (CDSA) is a test that checks for bacterial balance and health, digestive function and dysbiosis. A CDSA is used to determine what types of bacteria are found and measures beneficial, possibly harmful and disease-producing microbes. It also checks to see levels of candida. If present, they are cultured to see if they grow and what agents will be most effective in eliminating them. In addition, the CDSA measures digestive function by determining how well a person can digest proteins, fats and carbohydrates, the level of pancreatic enzymes produced, and the amount of short-chain fatty acids and butyric acid in the colon. Some labs also include a dysbiosis index, which uses the combined testing to give a measure of normalcy or abnormalcy. The CDSA provides a quick reference for your health care provider and can determine your therapeutic needs.

The CDSA test and associated research are very new. As recently as 1990, no one really knew what specific bacterial balance was normal, but since then, many thousands of stool analyses have been completed and the view of what is normal has become better defined. CDSA is useful for everyone who has digestive problems. If you only choose to do one test, do this one in conjunction with the comprehensive parasitology screening described below.

A recent client had ulcerative colitis characterized by bleeding with bowel movements, severe reactions to a large number of foods, rapid weight loss and fatigue. In partnership with his gastroenterologist we ran a CDSA. The test showed that he had overgrowth of two dysbiotic bacteria (citrobacter and klebsiella) and *Candida parapsilosis* (a less common variety of candida). Because lab technicians actually culture the client's microbes to see what works most effectively to kill them, the CDSA lists which medications the bacteria and fungi respond to most effectively, which the physician prescribed. In addition, we set up a dietary regime that eliminated foods which were irritating his digestive system. With proper diet, supplementation and medication, his bleeding stopped, he had normal bowel movements, he gained weight consistently, and is no longer taking prednisone. His physician had expected that he would need it for the rest of his life. The CDSA was validating to my client. He could finally "see" what was going on inside his body, which helped his self-esteem and his healing process tremendously. Because he has gotten so much better, he now has an optimistic view of his future health prospects and quality of life.

Parasitology Testing

Though we think of parasites as something we get from traveling in other countries, it's not true. According to the June 27, 1978, *Miami Herald*, the Center for Disease Control (CDC) in Atlanta found that 1 out of 6 randomly selected people had one or more parasites. Great Smokies Diagnostics Laboratory in North Carolina has similar results. They find parasites in 20 percent of samples tested. Over 130 types of parasites have been found in Americans.[1] Parasites have become more prevalent for many reasons, including contaminated water supplies, day care centers, ease of international travel, foods, increased immigration, pets and the sexual revolution. Most people will meet a parasite at some point in their lives. Contrary to popular myths, having parasites isn't a reflection of your cleanliness. My family contracted *Giardia* in Chicago. We hadn't traveled and have absolutely no idea how we contracted it.

If you have prolonged digestive symptoms, you should really consider having a comprehensive parasitiology screening. Some symptoms of parasites can appear to be like other digestive problems: abdominal pain, allergy, anemia, bloating, bloody stools, chronic fatigue, constipation, coughing, diarrhea, gas, granulomas, irritable bowel syndrome, itching, joint and muscle aches, nervousness, pain, poor immune response, rashes, sleep disturbances, teeth grinding, unexplained fever and unexplained weight loss.

Many physicians request parasitology testing on random stool samples. This can be highly inaccurate, so repeated testing is often necessary to get definitive results. Since many parasites live further up the digestive tract, many labs now give an oral laxative to induce diarrhea to detect these parasites. Others are found by using a rectal swab rather than a stool sample. The most accurate stool testing is usually done by labs that specialize in parasitology testing.

PARASITE QUESTIONNAIRE[2]

Check If Yes

1. Have you ever been to Africa, Asia, Central or South America, China, Europe, Israel, Mexico or Russia? ______
2. Have you traveled to the Bahamas, the Caribbean, Hawaii or other tropical islands? ______
3. Do you frequently swim in freshwater lakes, ponds or streams while abroad? ______
4. Did you serve overseas while in the military? ______
5. Were you a prisoner of war in World War II, Korea or Vietnam? ______
6. Have you had an elevated white blood count, intestinal problems, night sweats or unexplained fever during or since traveling abroad? ______
7. Is your water supply from a mountainous area? ______
8. Do you drink untested water? ______
9. Have you ever drunk water from lakes, rivers or streams on hiking or camping trips without first boiling or filtering it? ______
10. Do you use plain tap water to clean your contact lenses? ______
11. Do you use regular tap water that is unfiltered for colonics or enemas? ______
12. Can you trace the onset of symptoms (intermittent constipation and diarrhea, muscle aches and pains, night sweats, unexplained eye ulcers) to any of the above? ______
13. Do you regularly eat unpeeled raw fruits and raw vegetables in salads? ______
14. Do you frequently eat in Armenian, Chinese, Ethiopian, Filipino, fish, Greek, Indian, Japanese, Korean, Mexican, Pakistani, Thai or vegetarian restaurants; in delicatessens, fast-food restaurants, steak houses or sushi or salad bars? ______
15. Do you use a microwave oven for cooking (as opposed to reheating) beef, fish or pork? ______

16. Do you prefer fish or meat that is undercooked, i.e., rare or medium rare? ______
17. Do you frequently eat hot dogs made from pork? ______
18. Do you enjoy raw fish dishes like Dutch green herring, Latin American ceviche or sushi and sashimi? ______
19. Do you enjoy raw meat dishes like Italian carpaccio, Middle Eastern kibbe or steak tartare? ______
20. At home, do you use the same cutting board for chicken, fish, and meat as you do for vegetables? ______
21. Do you prepare gefilte fish at home? ______
22. Can you trace the onset of symptoms (anemia, bloating, distended belly, weight loss) to any of the above? ______
23. Have you gotten a puppy recently? ______
24. Have you lived with, or do you currently live with, or frequently handle pets? ______
25. Do you forget to wash your hands after petting or cleaning up after your animals and before eating? ______
26. Does your pet sleep with you in bed? ______
27. Does your pet eat off your plates? ______
28. Do you clean your cat's litter box? ______
29. Do you keep your pets in the yard where children play? ______
30. Can you trace the onset of your symptoms (abdominal pain, distended belly in children, high white blood count, unexplained fever) to any of the above? ______
31. Do you work in a hospital? ______
32. Do you work in an experimental laboratory, pet shop, veterinary clinic or zoo? ______
33. Do you work with or around animals? ______
34. Do you work in a daycare center? ______
35. Do you garden or work in a yard to which cats and dogs have access? ______
36. Do you work in sanitation? ______

37. Can you trace the onset of symptoms (gastrointestinal disorders) to any of the above? ______
38. Do you engage in oral sex? ______
39. Do you practice anal intercourse without the use of a condom? ______
40. Have you had sexual relations with a foreign-born individual? ______
41. Can you trace the onset of symptoms (persistent reproductive organ problems) to any of the above? ______

Major Symptoms: Please note that although some or all of these major symptoms can occur in any adult, child or infant with parasite-based illness, these symptoms might instead be the result of one of many other illnesses.

ADULTS

1. Do you have a bluish cast around your lips? ______
2. Is your abdomen distended no matter what you eat? ______
3. Are there dark circles around or under your eyes? ______
4. Do you have a history of allergy? ______
5. Do you suffer from intermittent diarrhea and constipation, intermittent loose and hard stools or chronic constipation? ______
6. Do you have persistent acne, anal itching, anemia, anorexia, bad breath, bloody stools, chronic fatigue, difficulty in breathing, edema, food sensitivities, itching, open ileocecal valve, pale skin, palpitations, PMS, puffy eyes, ringing of the ears, sinus congestion, skin eruptions, vague abdominal discomfort or vertigo? ______
7. Do you grind your teeth? ______
8. Are you experiencing craving for sugar, depression, disorientation, insomnia, lethargy, loss of appetite, moodiness or weight loss or gain? ______

CHILDREN

1. Does your child have dark circles under his or her eyes? ______
2. Is your child hyperactive? ______
3. Does your child grind or clench his teeth at night? ______
4. Does your child constantly pick her nose or scratch her behind? ______
5. Does your child have a habit of eating dirt? ______
6. Does your child wet his bed? ______
7. Is your child often restless at night? ______
8. Does your child cry often or for no reason? ______
9. Does your child tear her hair out? ______
10. Does your child have a limp that orthopedic treatment has not helped? ______
11. Does your child have a brassy, staccato-type cough? ______
12. Does your child have convulsions or an abnormal electroencephalogram (EEG)? ______
13. Does your child have recurring headaches? ______
14. Is your child unusually sensitive to light and prone to blinking frequently, eyelid twitching or squinting? ______
15. Does your child have unusual tendencies to bleed in the gums, the nose or the rectum? ______

INFANTS

1. Does your baby have severe intermittent colic? ______
2. Does your baby persistently bang his or her head against the crib? ______
3. Is your baby a chronic crier? ______
4. Does your baby show a blotchy rash around the peri-anal area? ______

Interpretation of Questionnaire

1. If you answered "yes" to more than 40 items, you are at high risk for parasites.
2. If you answered "yes" to more than 30 items, you are at moderate risk for parasites.
3. If you answered "yes" to more than 20 items, you are at risk.
4. If you are not exhibiting any overt symptoms now, remember that many parasitic infections can be dormant and then spring to life when you least expect them. Be aware that symptoms that come and go may still point to an underlying parasitic infection because of reproductive cycles. The various developmental stages of parasites often produce a variety of metabolic toxins and mechanical irritations in several areas of the body—for example, pinworms can stimulate asthmatic attacks because of their movement into the upper respiratory tract.

Food and Environmental Sensitivity Testing

Elimination/provocation testing for food and environmental irritants is a two-stage process: an elimination diet and then a provocation challenge test. Elimination/provocation testing requires your determination and commitment because you must radically change your diet to find out which foods you are reacting to. Often the foods we are sensitive to are the ones we depend on most, so you may have to do without some of your favorite foods for a while.

Over the past few years, I have been using an elimination diet that works well for nearly all of my clients (except those with candida). This program allows all fruit (except citrus), all vegetables (except tomatoes, eggplant, potatoes and peppers) and white rice. You can eat as much of these foods as you like, plus olive oil and canola oil for stir-frying and on salads. In addition, I use a rice-protein, nutrient-enriched drink that helps detoxify your system and ensures your protein needs are met. The foods allowed during the elimination are unlikely to cause food sensitivities. If food sensitivities are provoking symptoms, you'll find you feel terrific eating this way. In fact, some people feel better than they have in years! Elimination of the offending foods gives your digestive system a chance to heal over several months' time because you are no longer irritating it.

After 7 to 14 days on the elimination diet, you begin the provocation challenge. By slowly reintroducing foods into your diet, you can test yourself for your reactions. Do you become sleepy 30 minutes after eating wheat? Does cheese give you diarrhea? Do you itch all over after eating oranges? Do your joints ache after eating tomatoes? Through careful observation, you can detect many foods you have become sensitive to. Once you become familiar with your body's reaction, you can identify sensitivities, though it may be necessary to test a food several times to be certain of your reaction.

To enhance this process, it is best to also have a blood test for IgG4 and IgE antibody reactions to determine food and environmental sensitivities. Some labs also include testing for IgA and IgM antibodies. While many foods may be unmasked during the elimination provocation challenge, others may remain hidden. If you suspect that chemicals, molds or pollens are causing problems, you should also be screened for them. (Labs offer these tests either separately or as part of a complete screening package.) Sensitivities are rated from normal to severe reactions. In addition to a detailed readout documenting your personal reactions, most laboratories also include a list of foods which contain hidden sources of the offending foods, a rotation menu and other educational material to help you in the healing process.

Leaky Gut Syndrome/Intestinal Permeability Testing

There are several ways to measure for leaky gut syndrome/intestinal permeability. The method that has rapidly become the recognized standard is the mannitol and lactulose test. Mannitol and lactulose are water-soluble sugar molecules that our bodies cannot metabolize or use. They come in differing sizes and weights and are absorbed into the bloodstream at different rates. Mannitol is easily absorbed into cells by people with healthy digestion, while lactulose has such a large molecular size that it is only slightly absorbed. A healthy test shows high levels of mannitol and low levels of lactulose. If large amounts of mannitol and lactulose are present, it indicates a leaky gut condition. If low levels of both sugars are found, it indicates general malabsorption of all nutrients. Low mannitol levels with high lactulose levels have been found in people with celiac disease, Crohn's disease and ulcerative colitis.

Your doctor can give you a test kit to collect urine samples. After collecting a random urine sample, you drink a mannitol/lactulose mixture and collect urine for 6 hours. The samples are then sent to the laboratory. This test is often done in conjunction with a CDSA or a parasitology test.

Secretory IgA Testing

Secretory IgA (sIgA), an immunoglobulin antibody, is found in saliva in the mouth, throughout the digestive tract and in mucous secretions throughout the body. We normally don't think of the digestive system as being part of the immune system, but sIgA provides our first line of defense against bacteria, food residue, fungus, parasites and viruses. By sitting on mucous membranes, it prevents invaders from attaching to them and neutralizes them. New technology has made it simple to measure sIgA levels throughout the body using saliva.[3]

Deficiency of sIgA is the most common immunodeficiency. Low levels of IgA make us more susceptible to infection and may be a fundamental cause of asthma, autoimmune diseases, candidiasis, celiac disease, chronic infections, food allergies and more. In other words, if the sentry isn't standing at the gate, anyone can come in! A study examining people with Crohn's disease or ulcerative colitis found that all of them had low levels of sIgA. It concluded that raising sIgA levels might eliminate inflammatory bowel disease.[4]

High levels of sIgA are found in people who have chronic infections and whose immune systems are overloaded. It often accompanies chronic viral infections like CMV, EBV and HIV. It has also been found in people with rare medical problems like Berger's nephropathy, dermatitis herpetiformis, gingivitis, hepatic glomerulonephritis, IgA neoplasms, parotitis and secretory anti-sperm Abs.

Lifestyle and nutritional factors can influence sIgA levels. A balanced lifestyle, which encompasses our ability to nurture ourselves, environmental considerations, exercise, good food choices and moderate stress levels, can increase sIgA levels. Choline, essential fatty acids, glutathione, glycine, phosphatidylcholine, phosphatidylethanolamine, quercitin, vitamin C and zinc are all required to maintain healthy sIgA levels. Detoxification programs and repair of intestinal mucosa help normalize sIgA. *Saccharomyces boulardii*, a nontoxic yeast often found in probiotic supplements, has been shown to raise sIgA levels. A recent study showed that visualization and relaxation techniques significantly increase sIgA levels.[5]

Heidelberg Capsule Test

The Heidelberg capsule test is a radio-telemetry test for functional hydrochloric acid (HCl) levels. It is a simple, effective technique to determine how much HC1 your stomach is producing. Normally produced by the parietal cells of the stomach, HC1 is necessary for the initial stage of protein digestion and the absorption of vitamin B-12 and many minerals. Common symptoms of low-stomach acidity include belching or burning sensation immediately after meals, bloating, a feeling that the food just sits in the stomach without digesting and an inability to eat more than small amounts at any one sitting. Poor HC1 levels have been associated with childhood asthma, chronic hepatitis, chronic hives, diabetes, eczema, gallbladder disease, lupus erythematosus, osteoporosis, rheumatoid arthritis, rosacea, under- and overactive thyroid conditions, vitiligo and weak adrenals. (A full list of symptoms and conditions is listed in the section on Hypochlorhydria in the Self-Care Section.) As we age, we produce less HCl. As many as half of all people over the age of 60 may be deficient in this important substance.

The Heidelberg test has proven to be accurate and sensitive in determining stomach pH levels. After you swallow a radio transmitter that's about the size of a B-complex vitamin, it measures the resting pH of your stomach, alternating with challenges of baking soda, which is very alkaline. By observing how well the stomach returns to an acid condition after administration of the baking soda, the physician can determine whether or not you produce adequate amounts of HCl.

FUNCTIONAL LIVER PROFILE TESTING

Liver profile testing is useful for determining how well you are able to handle toxic substances. The liver is responsible for transforming toxic substances into harmless byproducts that the body can excrete. It does this in a variety of ways, including acetylation, conjugation, sulfation, sulfur transferase and via the cytochrome P450 system. The functional liver profile looks at the total toxic load in your body and sees how well your body handles it by measuring the abilities of your liver detoxification systems.

When we are exposed to substances that the body sees as toxic, cytochrome P450 levels go up. Many substances can stimulate its production, including alcohol, barbiturates, carbon tetrachloride, charcoal broiled meats, dioxin, exhaust fumes, high-protein diets, niacin, oranges, organophosphorus pesticides, paint fumes, riboflavin, sassafras, saturated fats, steroid hormones, sulfonamides and tangerines.[6] The cytochrome P450 system transforms endotoxins (toxins produced within the body) and exotoxins (those taken into the body from outside) into water-soluble forms that can be excreted in urine. By measuring the cytochrome P450 levels and seeing how quickly the toxic materials are transformed, tests can provide useful information about how well the body is able to detoxify a wide variety of substances.

When cytochrome P450 is released, a second phase of detoxification is also activated, which causes an elevation of D-glucaric acid. D-glucaric acid is released by the liver and is a byproduct of turning toxic substances into glucuronidase, which is excreted in urine. A sulfur/creatinine ratio is used to test this function. This ratio helps determine leaky gut syndrome, low glutathione levels and the level of free radical activity.

For the test you drink a beverage which contains caffeine and sodium benzoate at home and send samples off to the laboratory for analysis. By measuring your urine and saliva, it can be determined how well your liver is able to process toxins. Normal levels of cytochrome P450 and benzoate are found in 50 percent of people tested. A low-caffeine clearance is found in about one-third of all people tested and indicates that your body is having difficulty detoxifying. High-caffeine clearance levels are found in people who have been exposed to high levels of toxins and smoke.

Slow benzoate conversion indicates that secondary detoxification is not working normally. When coupled with a high cytochrome P450 level, it indicates that high levels of toxic substances are being transformed into harmless forms too slowly. This occurs in about 17 percent of people tested. They experience an aggravation of their symptoms when on a detoxification program and must be carefully monitored. These people benefit from a nutrient-rich program to build them up before beginning a detox program.[7]

Another detoxification pathway is the conversion of glycine to hippuric acid. This is the reaction by which the toxic substances are converted into water-soluble forms to be excreted. A sodium benzoate drink—sodium benzoate is a common food preservative found in pickles—is used to test for glycine conjugation. Sodium benzoate gets changed into hippuric acid and is then measured in the urine.

The liver function profile also checks sulfate and creatinine levels. Low sulfate to creatinine ratios indicate poor levels of glutathione and sulfate that are essential for proper detoxification.

INDICAN TEST

The Indican test measures the amount of putrefaction in your digestive system. This is a urinary test and gives a general indication of how well you are digesting foods. High indican levels are found in people with dysbiotic conditions and malabsorption. Indican testing offers a quick way to screen for faulty digestion, but does not give enough detailed information to know exactly why or where the problem originates. Some doctors find it a useful office screening test because it is inexpensive and noninvasive.

Lactose Intolerance

Lactase is an enzyme that digests lactose, a sugar naturally found in milk products. The inability to digest lactose affects about 70 percent of the world's population and is highly prevalent in African Americans, Asian Americans, Caucasian Americans of Mediterranean and Jewish descent, Hispanics and Native Americans. Lactose intolerance is caused by an enzymatic deficiency and is not a milk allergy, which is the inability to digest milk proteins. Lactose intolerance causes a wide variety of symptoms including abdominal cramping, acne, bloating, diarrhea, gas, headaches and nausea. Most people with lactose intolerance fail to recognize that the food they eat has any relationship to how they feel.

There are two ways to test for lactose intolerance—a self-test and a laboratory test. To self-test, you must restrict intake of all dairy products for at least ten days. Obvious sources include milk, yogurt, cheese, ice cream, creamed soups, frozen yogurt, powdered milk and whipped cream. Less obvious sources are bakery items, cookies, hot dogs, lunchmeats, milk chocolate, most nondairy creamers, pancakes, protein powder drinks, ranch dressings and anything that contains casein, caseinate, lactose, sodium caseinate and whey. If you're not sure what's in a food, avoid it during the testing period. It's probably best to eat all your meals at home or prepare all food yourself.

If lactose intolerance is causing your problems, you will probably notice your symptoms have disappeared significantly. Reintroduction of dairy products will trigger a return of symptoms. However, the results may be inconclusive because you may be sensitive to other foods in addition to dairy—in which case your symptoms won't change. A laboratory test would then be indicated.

Your doctor can order a simple, noninvasive hydrogen breath test to pinpoint if lactose intolerance is causing your problems. This challenge test is ideal for people who find it difficult to complete a self-test or are confused about their findings. After you breathe into a bag to collect a baseline sample, you drink a small amount of a lactose solution and breathe into a different bag. Lab technicians measure the levels of hydrogen and methane gas you exhaled in both samples. Normal methane levels are 0 to 7 ppm. If levels increase at least 12 ppm between the two samples, it indicates lactose intolerance, even if your hydrogen production is normal. Normal hydrogen levels are 10 ppm. Levels of 20 ppm or more are commonly found in people with lactose intolerance. When both methane and hydrogen are measured, false results are narrowed considerably. Use of antibiotics, enemas and laxatives are common reasons for false negative results, which occur 5 percent of the time. This test has a few false positive results, which are generally caused by eating high-fiber foods before the test, exposure to cigarette smoke or sleeping during testing.

Hair Analysis

Hair analysis is a useful tool to see how well you absorb essential minerals and what levels of toxic minerals have accumulated in your tissues. Hair is cut as close to your head as possible—only the 1 to 2 inches which grew closest to your head will be tested because it reflects your mineral status over the past few months. The lab burns your hair in an electrochromatography scan to measure the levels of minerals present. Low levels of six or more essential minerals indicate a problem with absorption of nutrients. This could be due to drug therapy, dysbiosis, low hydrochloric acid levels or poor flora. Hair analysis is also an accurate test to determine whether you have high levels of toxic minerals like aluminum, arsenic, cadmium, lead and mercury. It is also a recognized tool for detecting lead poisoning in children.

pH Testing

Our body pH is maintained just above 7.0 for optimal health. Note that 7.0 is the pH of water, which is neither alkaline nor acidic. pH is kept in balance by a variety of mechanisms which produce acid byproducts. The alkaline mineral buffers, found in fruits and vegetables, help keep us healthy. Many disease conditions are helped by keeping pH at optimal levels.

pH testing is a simple home urine test with a dip stick. You simply place a piece of pH-sensitive paper into your first morning urine and "read" it to see what your pH is. Optimally, the level will be between 6.5 and 7.5, or fairly neutral. By manipulating your foods, you can bring this into an optimal range, which will give support to your body's own innate healing and balancing capacities.

Small Bowel Bacterial Overgrowth

The small bowel bacterial overgrowth test measures breath levels of hydrogen and methane to determine if there is a bacterial infection in the small intestine. This test differs from the comprehensive digestive and stool analysis in that it tests for dysbiosis of the small rather than the large intestine. Small bowel overgrowth occurs when bacteria in the large intestine travel to the small intestine. It is often the result of poor hydrochloric acid production in the stomach or an insufficient amount of pancreatic enzyme function. It is often found in conjunction with parasitic infections. Chronic pancreatitis, Crohn's disease and lupus erythematosus can also cause small bowel overgrowth.[8] People with small bowel bacterial overgrowth experience diarrhea, poor nutrient absorption and weight loss. Other people who may be affected are those with poor ileocecal valve function, poor intestinal motility, scleroderma or recent gastric surgery. People in-

FOOD & CHEMICAL EFFECTS ON

Most Alkaline	More Alkaline	Low Alkaline	Lowest Alkaline
•Baking Soda	Spices/ Cinnamon	Herbs (most)	
Table salt (NaCl)	Sea salt		*Sulfite*
Mineral water	•Kambucha	•Green or mu tea	Ginger tea
	Molasses	Rice syrup	•Sucanat
	Soy sauce	Apple cider vinegar	•Umeboshi vinegar
•Umeboshi plums		•Sake	•Algae,blue-green
			•Ghee (clarified butter)
		•Quail eggs	•Duck eggs
			Oats
			'Grain coffee'
			•Quinoa
			Wild rice
	Poppy seed	Primrose oil	Avocado oil
Pumpkin seed		Sesame seed	Seeds (most)
	Chestnuts	Cod Liver oil	Coconut oil
	Pepper	Almonds	Olive oil
Hydrogenated oil		•Sprouts	Linseed oil
Lentils	Kohlrabi	Potato/bell pepper	Brussel sprout
Yam	Parsnip/Taro	Mushroom/fungi	Beet
	Garlic	Cauliflower	Chive/cilantro
Onion	Kale/Parsley	Rutabaga	Okra
Daikon/•Taro root	Endive	•Salsify/•ginseng	Turnip greens
Sea vegetables (other)	Mustard green	Eggplant	Squashes
Burdock/Lotus Root	Ginger root	Pumpkin	Lettuces
Sweet potato/yam	Broccoli	Collard greens	Jicama
Lime	Grapefruit	Lemon	Orange
Nectarine	Canteloupe	Pear	Apricot
Persimmon	Honeydew	Avocado	Banana
Raspberry	Citrus	Apple	Blueberry
Watermelon	Olive	Blackberry	Pineapple
Tangerine	•Dewberry	Cherry	Raisin, currant
Pineapple	Loganberry	Peach	Grape
	Mango	Papaya	Strawberry

•Therapeutic, gourmet or exotic items. *Italicized items are NOT recommended.*

Used with permission of Russell Jaffe, M.D., *Health Collegieum.*

ACID/ALKALINE BODY CHEMICAL BALANCE

Food Category	Lowest Acid	Low Acid	More Acid	Most Acid
Spices/Herbs	Curry	Vanilla	Nutmeg	Pudding/Jam/Jelly
Preservatives	*MSG*	*Benzoate*	*Aspartame*	
Beverages	*Kona coffee*	*Alcohol*	Coffee	Beer
		Black tea		Yeast/hops/malt
Sweeteners	Honey/maple syrup		*Saccharin*	Sugar/cocoa
Vinegar	Rice vinegar	Balsamic vinegar		White/acetic vinegar
Therapeutics		*Antihistamines*	*Psychotropics*	*Antibiotics*
Processed dairy	Cream	Cow milk	•Casein, milk protein	*Processed cheese*
Cow/human soy	Yogurt	Aged cheese	New cheeses	Ice cream
		Soy cheese	Soy milk	
Goat/sheep	Goat/sheep/cheese	Goat milk		
Eggs	Chicken eggs	•Game meat		
Meat	Gelatin/organs	Lamb/mutton	Pork/veal	Beef
Game	venison	Boar/elk	Bear	
Fish/shell fish	Fish	Shell fish	•Mussles/squid	Lobster
Fowl	Wild duck	Goose/turkey	Chicken	•Pheasant
	Triticale	Buckwheat	Maize	Barley
Grains	Millet	Wheat	Barley groats	
Cereal	Kasha	•Spelt/teff	Corn	
Grass	Amaranth	Farina/ semolina	Rye	
	Brown rice	White rice	Oat bran	
Nuts	Pumpkin seed oil	Almond oil	Pistachio seed	•*Cottonseed oil/meal*
Seeds/sprouts	Grape seed oil	Sesame oil	Chestnut oil	Hazelnuts
Oils	Sunflower oil	Safflower oil	*Lard*	Walnuts
	Pine nuts	Tapioca	Pecans	Brazil nuts
	Canola oil	•Seitan	Palm kernel oil	*Fried foods*
	Spinach	Tofu	Green pea	Soybean
Beans	Fava beans	Pinto beans	Peanut	Carob
Vegetables	Kidney beans	White beans	Snow pea	
Legume	String/wax	Navy/red	Legumes (other)	
Pulses		Azuki beans	Carrots	
Roots	Chutney	Lima beans	Chickpeas	
	Rhubarb	Chard		
Citrus fruits				
	Guava	Plum	Cranberry	
	•Pickled fruit	Prune	Pomegranate	
	Dry fruit	Tomatoes		
Fruits	Figs			
	Persimmon juice			
	•Cherimoya			
	Dates			

fected with small bowel overgrowth often have difficulty with digestion of fats, which come through undigested in the stool, called steatorrhea. They may also experience B-12 deficiency, chronic diarrhea and poor absorption of the fat-soluble vitamins A, D, E and K.[9]

Breath testing provides a simple, noninvasive alternative to the more widely used method of obtaining a small bowel aspirate and is more accurate.[10] To perform the test, you drink either a lactulose or a glucose drink and collect breath samples. Hydrogen is produced when lactulose or glucose come in contact with the gut flora. A significant rise in hydrogen levels indicates small bowel overgrowth.

EAV TESTING

The electrical acupuncture voltage (EAV) test is widely used in Europe after much positive research. Although it has met with FDA resistance in the United States, there are many skillful professionals who use this test to successfully diagnose and determine appropriate therapies.

The test measures the electrical activity of your skin at designated acupuncture points. You hold a negative rod in one hand, the practitioner places a positively charged pointer on a variety of points on your skin, and a meter measures the voltage reading between the points. The test can determine which organs are strong or weak, which foods help or hurt you, which nutrients you need or have excessive levels of, and how old patterns are contributing to your health today. It is a fast, noninvasive screening test.

Part 4

The Digestive Wellness Program:

You Can Change the Way You Feel!

Chapter 8

Moving Toward a Wellness Lifestyle

"Health approaches seek to identify the causes (or potential causes) of disease and employ therapies designed to eliminate host susceptibility and hospitality to ill health. Health approaches give priority to therapies of minimum risk and high therapeutic gain, tailored to the individual's attitudes and circumstances."

Russ Jaffe/Women and Children's Health Update, 1994-5 Syllabus

Health is a continuum. At one end you have optimal health and at the other death.

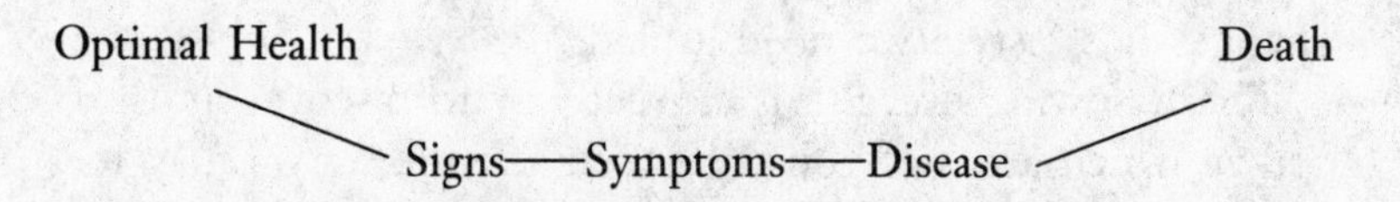

Somewhere to the right of optimal health, you find you have less energy than you would like to have. Maybe you

experience discomfort after eating. If at this point you contact your physician, he/she will probably tell you to chew your food more thoroughly and take an over-the-counter medication for discomfort. You may find over time the problem disappears. The body's own innate healing balance has restored itself without much help from you. On the other hand, you may find that the discomfort is getting worse. Now you also experience bloating and gas. Again, your physician is likely to recommend a simple medication. And over time, the problem may go away. If it continues to get worse, your physician will be able to make a "diagnosis" and give a clinical name to your problem—yes, you have diverticulitis or irritable bowel syndrome.

In a wellness scenario, the time to tackle the problem is when you have the first small signs. Paying attention to what your body is trying to tell you is the first principle of wellness. Indigestion is not caused by a Rolaids deficiency. Listening to your body can help you make healthful changes in your lifestyle that will affect not only your tummy, but your entire well-being. For example, if your indigestion happens when your stress level is high, then you can focus on building stress management tools. If your indigestion happens after you eat milk products, then the solution may be the elimination of dairy products. If your indigestion is worse when you are out of shape, then an exercise program may be in order.

Most of us make short-lived attempts toward wellness. We begin a new diet each Monday, we make New Year's resolutions that we've forgotten by Valentine's Day and other plans get washed away by daily responsibilities. It's not that we aren't trying, we are. But in order to implement long-lasting changes in lifestyle we need a plan.

The following exercise is a great way to discover and prioritize which areas of your life need attention. You may feel your relationships have little or nothing to do with the fact that you have chronic diarrhea, but until you balance your relationships, you can't know for sure. The mind is not separate from the body. It is well-documented that the thoughts we have influence our physical condition. All domains of wellness affect our sense of well-being.

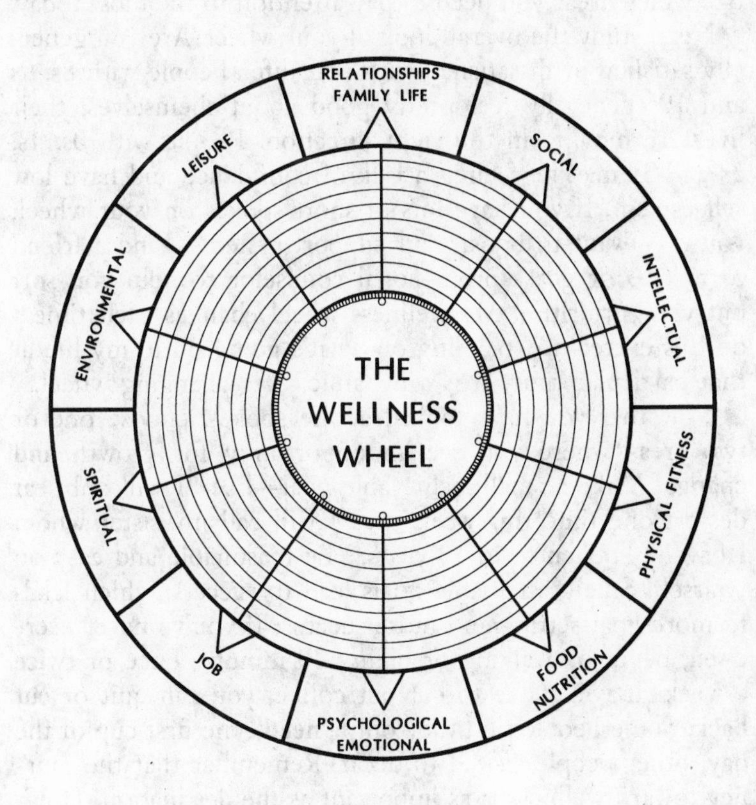

Wellness Wheel[1]

How satisfied are you today in each of these areas of your life? Rate them on a scale from 1 to 10, with 1 being the most dissatisfied to 10 being the most satisfied.

Now look at your wellness wheel. Would it roll? Which

spokes are the shortest? The short spokes are the areas you've paid less attention to recently, so they offer the greatest areas of discovery and opportunity. They help you prioritize which areas you need to pay attention to the most today.

Next, study the overall look of your wheel. Are you generally satisfied or dissatisfied with your life? People with 8s, 9s and 10s generally feel pretty good about themselves; their lives are moving in the right direction. People with 0s, 1s, 2s and 3s may be feeling a lack of confidence and have low self-esteem. If you have lots of short spokes on your wheel, you may want to boost your support systems. Find a friend to talk to or get a professional counselor to help you sort out your priorities. My wellness wheel changes each time I do this exercise, depending on what's going on in my life at that particular time. It is a dynamic, ever-changing wheel.

Now that you have looked at the spokes, choose one or two areas where you see an opportunity for growth and change. Make a small, achievable goal—like "I will only eat dessert one time this week" or "I will call my sister whom I haven't spoken to in 20 years." Be reasonable and easy on yourself. Small, attainable goals lead to success, which leads to more goal-setting and more success. If you've never exercised, begin by walking or biking 20 minutes once or twice a week. If you'd like to cut out coffee, you can quit or cut back. Some people find they only "need" the first cup of the day; other people switch to decaf. Remember that the journey toward wellness is as important as the destination. Don't worry about "getting there"; just enjoy the scenery!

When you work consistently on a wellness lifestyle, you expect to be healthier in ten years than you are today. Of course, we do age and we may meet with an unexpected illness, but in general, people who pay attention to how they feel have a greater sense of well-being. They also feel more responsible and in control of their lives.

In our culture, we approach aging as a "going-down-hill" process. We expect to spend the last 20 to 30 years of our lives adapting to loss of function. But we all have friends who lead active, healthy lives until ripe old age. I have a friend who is 96 years old, works as a stock broker 4 days a week and spends a long weekend at his country home. His health isn't perfect, but his life is full and his mind is interested and busy. His body is aging, but his spirit is youthful. Anyone who wants to can change their expectations and set goals of greater healthspan and well-being.

Wellness is more than the absence of illness. Wellness is having the energy to do the things you dream of. Wellness is the belief that your body wants to heal itself and that you can improve your health over time, rather than stand by and watch it deteriorate. Does that mean you won't age? No, all people age. But we can build a strong house, so that we age gracefully. Having a healthy body prevents chronic, degenerative illnesses that are primarily products of our lifestyle. Wellness demands that we take responsibility for our own health and make changes in the way we live in order to help our body function more optimally. We are required to ask questions, rather than just accept the cultural norm.

According to a study published in the *New England Journal of Medicine*, 30 percent of all Americans paid for "alternative" health care in 1989. These visits were paid by the person and were in addition to the regular care medical physicians provide. We are in a transitional period as we recognize that Western medicine only offers one slice of the health care pie. Other healing modalities offer valid and complementary slices which complete the pie.

In the following chapters, you will be given a wellness prescription which includes general guidelines, followed by specific guidelines. Because you are unique, you need to tailor these plans to suit your current needs and lifestyle. You would benefit from a consultation with a Certified Clinical Nutritionist (CCN), a Certified Nutrition Specialist (CNS) or a physician who specializes in nutritional medicine. These professionals can help you determine the essentials of your program, its duration and how to work with it over time. But remember you are the boss. You're the only one who knows exactly how you feel.

When you find a health professional whom you like, trust, respect and are attuned with, listen to his or her advice. When you agree on a plan of action, follow through. You can only really tell what the results will be if you follow the plan over a period of time. If you feel that a given therapy isn't helping after a set amount of time, it may not be the correct one for you. If you have a bad feeling about a therapy or about the person you are working with, do something about it. Be bold, ask questions, educate yourself and direct your own care.

Many people "shop around" for therapeutics and practitioners and never really complete any given program. You aren't doing yourself a favor if you don't make a commitment. If you want to add another therapy or modality to your program, talk it over with the health professional you are working with. Team efforts often give the best results, but only with commitment and communication.

Here are some people you may want to put on your wellness team: medical doctor, chiropractor, naturopathic physician, clinical nutritionist, physical therapist, acupuncture practitioner, herbalist, homeopath, psychologist, colon therapist, massage therapist. Each of these professionals can guide you toward health from their own perspective. There are many roads to wellness. If you find yourself at a dead end, turn around and take another road.

Chapter 9

First Things First: Detoxification

"It can be strongly said that the health of an individual is largely determined by the ability of the body to detoxify."

Joseph Pizzorno, Michael Murray, N.D.s,
Encyclopedia of Natural Health,
Prima Publications, 1991

Cleansing and fasting are integral to holistic healing. The basic tenets of functional medicine are clean the body, remove any irritants, whether they be food, medicine or microbes, provide the proper nutrients for the body to use as building materials, replace missing intestinal flora, enzymes and hydrochloric acid and give the body time to heal itself.

We are exposed to toxins everywhere—from the air we breathe to the foods we eat, even as a result of metabolism. These toxins cause irritation and inflammation throughout our bodies. People have always been exposed to toxic substances, but today's exposure to contaminants far exceeds that of previous times. Each week approximately 6,000 new chemicals are listed in the Chemical Society's *Chemical Abstracts*, which adds up to over 300,000 new chemicals each year. On average we consume 14 pounds each year of food additives, including colorings, preservatives, flavorings, emulsifiers, humectants and antimicrobioals.[1] In 1990 the EPA estimated there were 70,000 chemicals commonly used in pesticides, foods and drugs. In 1992 the National Research Council published a report which suggested that environmental toxins play a role in neurological illnesses like Parkinson's disease, Alzheimer's disease and amyotrophic lateral sclerosis.[2] The link between breast cancer and pesticide usage is becoming stronger and stronger. It is estimated that up to 25 percent of the U.S. population has heavy metal poisoning.

Our body normally produces toxins as a byproduct of metabolism. We call these toxins endotoxins, which means they come from within us. If not eliminated, these endotoxins can irritate and inflame our tissues, blocking normal functions. Endotoxins formed by bacteria and yeasts can be absorbed into the bloodstream. Antibodies formed to protect us against the harmful effects of these endotoxins often trigger a systemic effect, causing an autoimmune reaction, so our body begins fighting itself.

By assisting your body in removal of stored toxins through detoxification programs, your body can more easily heal itself. Although cleansing is important for healthy people, it is essential for people who don't feel as well as they would like to. One of the many functions of the liver is to act as a filter, to let nutrients pass, if possible to "humanize" other substances and to transform toxins into safe substances that can be eliminated in urine and stools. When the liver enzymes fail to break down these toxins, they are stored in the liver and fatty tissue throughout our bodies. Common medications can inhibit the liver's ability to adequately process toxins. Acetaminophen (Tylenol) causes liver damage when used in combination with alcoholic beverages. Cimetidine, an ulcer medication, limits the liver's ability to detoxify foreign substances. A thorough cleansing program works systemically, cleansing all the cells in your body of harmful toxins.

Throughout time and in various cultures, people have seen the need for periodic internal cleansing. Native Americans use sweat lodges. Ancient Roman bath houses had rooms for bathing in steam, warm water and cold water. Jewish women have used ritual mikva baths to cleanse both body and spirit. Most Swedish people have home saunas, and our own health clubs have saunas, streambaths and Jacuzzis. People "take the waters" in Europe and parts of the United States. Hawaiians use steam and a form of massage, called *lomilomi*, where they scrub people clean with the red Hawaiian dirt and sea salt. In fact, mud and clay have been used worldwide to draw toxins from the body while simultaneously providing essential nutrients.

Fasting is an important part of many religious holidays and customs. Both Jesus and John the Baptist fasted to gain mental and spiritual clarity. During Ramadan, an important Muslim holiday, people fast during daylight hours for a month. Jewish people fast on Yom Kippur. Indigenous people of many cultures use fasting as a way to clarify thought and provoke vision.

Removal of waste material—detoxification— is essential to the healthy functioning of our bodies. This is shown in the many different ways the body cleans itself. Although we typically give little thought to it, skin is our body's largest organ. In addition to being a protective organ, it is also an organ of elimination through perspiration. Sneezes clear our sinuses. Lungs breathe out carbon dioxide, and even the breath allows for removal of some wastes. Kidneys filter wastes from the bloodstream. Stool is the residue from the digestive process. The liver filters the substances that are absorbed through the digestive barrier into the bloodstream. White blood cells gobble up bacteria and foreign substances, and the lymphatic system clears the debris from circulation. During a cleansing program, your body more rapidly recycles materials to build new cells, take apart aged cells and repair damaged cells.

Most detoxification programs focus on the liver. The liver is, in my opinion, the most overworked organ of the body. It has responsibility for manufacturing 13,000 different enzymes, producing cholesterol, breaking down estrogen, regulating blood sugar, filtering blood, manufacturing bile, breaking down old red blood cells and detoxifying harmful substances. When the liver loses its ability to easily perform these functions, we begin to feel poorly, with many systems out of balance.

To enhance the liver's ability to perform, it is essential to flush out all unnecessary substances. The liver changes toxic substances by converting them from a fat soluble form into a mycellized form that the kidneys can excrete. This process occurs in the cytochrome P450 pathway, also known as phase I deconjugation. When this pathway becomes overloaded, the liver loses its conversion capacity and substances get stored in the liver and in fat cells throughout the body. In phase II the body removes these substances from storage. (Functional liver testing can actually show how well your phase I and phase II detoxification pathways are working.) These toxins affect how we feel, contributing to poor immune function, low energy levels, lack of digestive wellness, poor functioning of heart and respiratory systems and overall lowered health status. Removal of these toxins can be transforming on many levels, providing an increase in energy and creativity.

It is best if you follow a detoxification program under the supervision of a medical or health professional. Toxins released too quickly can make you feel worse than when you began and can aggravate your symptoms. For instance, a detox program may provoke temporary symptoms of fatigue, headaches, diarrhea, constipation, irritability and lightheadedness, which usually subside within the first few days. A professional can guide you gently through the process so that you get the maximum benefits for both mind and body with minimum discomfort.

DETOXIFICATION PROGRAMS

There are many different detoxification programs. Fasting, modified fasting, metabolic cleansing, colonic irrigation, steaming, mud packs, saunas, herbal detoxification programs and hot tubs all have therapeutic detoxification benefits. When choosing a detoxification program, it must meet specific criteria: It needs to (1) work with your life and your values, (2) be thorough, and (3) be gentle and nurturing to your body.

Over the past several years, I have personally and professionally relied on three main detoxification programs that are effective and gentle: fruit and vegetable cleansing, metabolic cleansing and low-temperature steams and saunas. I also recommend vitamin C flushes between cleansings. Other professionals may prefer fasting programs or colonic irrigation, which in the right hands can be powerful tools for healing.[3] Because there are many fine books on fasting and colonic irrigation, I have not included information about them here.

Fruit and Vegetable Cleansing

This gentle cleansing method is outlined in Chapter 6 as the elimination portion of the elimination provocation challenge. You eat all you want of fruits, vegetables and rice and use olive and canola oils as condiments for 7 to 10 days. Fresh fruit and vegetables juices are an excellent source of easily assimilated nutrients and alkalizing minerals and can enhance the detoxification pathways. It's important to eat every two to three hours to keep your blood sugar levels normal. The major benefit of this detox method is that you can do this on your own without professional supervision. Of course, if you are under a doctor's care or taking medication of any kind, you'll need to let your physician know of your plans. The first few days may require mental and physical adjustments to your new regimen, but most people feel a sense of general well-being. You may notice that many of your outstanding symptoms have disappeared or become less aggravated.

You may also experience some discomfort during the first three or four days. Headaches are fairly common and may be the result of withdrawal from caffeine, sugar, alcohol or other substances. They are an indicator that toxins are being flushed out or that your body is going through withdrawal. To facilitate this, drink a lot of water, diluted juices and all herbal teas except those containing caffeine. Some people develop rashes or pimples as the skin works hard to eliminate toxins. Saunas, steambathing and massaging your skin with a soft, dry brush or loofa can help your skin. If you are constipated, make sure you eat enough fiber-rich fruits and vegetables (apples, broccoli, pears, sweet potatoes, peas, brussel sprouts, corn, potatoes, carrots, greens, blackberries, bananas, strawberries, raspberries, spinach). If you are constipated, toxins may be reabsorbed into your bloodstream, causing symptoms such as headaches and nausea. So add a fiber supplement; psyllium seeds or psyllium seed husks work well. Begin with 1 teaspoon in water, and drink up quickly before it turns into a gel. Aloe vera juice may also help regulate your bowels.

Metabolic Cleansing

Metabolic cleansing is a gentle yet deep method of detoxification, and it is the best, most thorough program I have used. The foundation of this program is a hypoallergenic/sensitive rice-based protein and nutrient drink. The first week you eat only the drink and all you want of rice, all fruits except citrus, all vegetables except nightshade family foods and small amounts of oils.[4] Citrus fruits are excluded because many people are sensitive to them. Nightshade vegetables—potatoes, tomatoes, eggplant, green peppers and chili peppers—are excluded because 10 to 15 percent of people with joint pain become painfree when these vegetables are removed from the diet. If you know that you are sensitive to bananas, melons or other foods that are allowed on the plan, then avoid them as well. It is possible for people to continue their normal daily routine while on this program, whereas juice fasting requires nearly total rest and relaxation.

This program is administered only through health professionals who can monitor your progress, determine when you should quit and help you adjust if you have any difficulties. This program may be inappropriate or may need modification for people in poor health or on specific medications and must be supervised by someone who is familiar with its effects. People who score over 50, or 10 or more in any one section on the Metabolic Screening Questionnaire below, may want to begin with repair and do a detoxification later. A different protein drink helps heal your digestive tract and can be used over time to strengthen your system before the detox.

During a metabolic cleansing most people find a dramatic alleviation of symptoms and a distinct improvement in energy level. The high levels of nutrients found in the rice protein drink and in the fruits and vegetables help the liver activate its detoxification pathways and move unwanted materials out of the body. The intention here is to allow your digestive system to rest, relax and heal itself.

Once you have completed the cleansing, it is important to slowly reintroduce foods back into your diet. This is an ideal time to screen yourself for food sensitivities, as described in Chapter 6. Being a food sleuth takes a lot of patience; it's not always apparent which foods cause adverse symptoms. But with persistent detective work you can discover many of them. Keep a running record of everything you eat and of your symptoms. Food sensitivities often display delayed reactions, so it may take up to 48 hours to feel the effect of a newly introduced food.

You may want to have testing done for food allergies and sensitivities at this time. If you have uncovered a problem, these tests can further pinpoint which foods and substances are making you ill.

Metabolic Screening Questionnaire[5]

Rate each of the following symptoms based on your health profile for the past 30 days.

POINT SCALE: ______________________________

0 = Never or almost never have the symptom.
1 = Occasionally have it; effect is not severe.
2 = Occasionally have it; effect is severe.
3 = Frequently have it; effect is not severe.
4 = Frequently have it; effect is severe.

DIGESTIVE TRACT	____ Nausea or vomiting	**Total** ____
	____ Diarrhea	
	____ Constipation	
	____ Bloated feeling	
	____ Belching or passing gas	
	____ Heartburn	
EARS	____ Itchy ears	**Total** ____
	____ Earaches, ear infections	
	____ Drainage from ear	
	____ Ringing in ears, hearing loss	

EMOTIONS	___ Mood swings	**Total** ___
	___ Anxiety, fear or nervousness	
	___ Anger, irritability or aggressiveness	
ENERGY/ ACTIVITY	___ Fatigue, sluggishness	**Total** ___
	___ Apathy, lethargy	
	___ Hyperactivity	
	___ Restlessness	
EYES	___ Watery or itchy eyes	**Total** ___
	___ Swollen, reddened or sticky eyelids	
	___ Bags or dark circles under eyes	
	___ Blurred or tunnel vision (Does not include near- or far-sightedness)	
HEAD	___ Headaches	**Total** ___
	___ Faintness	
	___ Dizziness	
	___ Insomnia	
HEART	___ Irregular or skipped heartbeat	**Total** ___
	___ Rapid or pounding heartbeat	
	___ Chest pain	
JOINTS/ MUSCLES	___ Pain or aches in joints	**Total** ___
	___ Arthritis	

	___ Stiffness or limitation of movement	
	___ Pain or aches in muscles	
	___ Feeling of weakness or tiredness	
LUNGS	___ Chest congestion	**Total** ___
	___ Asthma, bronchitis	
	___ Shortness of breath	
MIND	___ Poor memory	**Total** ___
	___ Confusion, poor comprehension	
	___ Poor concentration	
	___ Difficulty in making decisions	
	___ Stuttering or stammering	
	___ Slurred speech	
	___ Learning disabilities	
MOUTH/ THROAT	___ Chronic coughing	**Total** ___
	___ Gagging, frequent need to clear throat	
	___ Sore throat, hoarseness, loss of voice	
	___ Swollen or discolored tongue, gums, lips	
	___ Canker sores	
NOSE	___ Stuffy nose	**Total** ___
	___ Sinus problems	

	___ Hay fever	
	___ Sneezing attacks	
	___ Excessive mucus formation	
SKIN	___ Acne	**Total** ___
	___ Hives, rashes or dry skin	
	___ Hair loss	
	___ Flushing or hot flashes	
	___ Excessive sweating	
WEIGHT	___ Binge eating/drinking	**Total** ___
	___ Craving certain foods	
	___ Excessive weight	
	___ Compulsive eating	
	___ Water retention	
	___ Underweight	
OTHER	___ Frequent illness	**Total** ___
	___ Frequent or urgent urination	
	___ Genital itch or discharge	

Vitamin C Flush

High levels of vitamin C help detoxify the body, rebalance intestinal flora and strengthen the immune system. The vitamin C flush can be used between metabolic cleansing therapies or at the first sign of a cold or infection. If your immune system is weak or you've been exposed to a lot of toxins, you may want to do a vitamin C flush once a week for a month or two. On days when you are not doing a vitamin C flush, take a minimum of 2,000 to 3,000 mg. Humans are one of the only animals that do not produce their own vitamin C so we need to replenish our supply daily.

Vitamin C is necessary for maintenance of the cytochrome P-450 detoxification pathways in the liver. It is used by physicians intravenously to help remove toxins from the body during chelation therapy. Vitamin C has been well researched for its ability to help detoxify bacterial toxins, drugs, environmental toxins and heavy metals from our bodies. Its gentle and potent detoxification counteracts and neutralizes the harmful effects of manufactured poisons.[6]

To do a vitamin C flush, you take vitamin C to the level of tissue saturation. You'll know you've reached it because you will have watery diarrhea. You'll need to purchase powdered mineral ascorbate C, which is more easily tolerated by most people because it doesn't change your pH balance. The amount you take varies depending on your personal needs that day. Many of us require about 5,000 mg; others need 15 or 20 times as much. For instance, if you're coming down with a cold, have chronic fatigue syndrome or are under excessive stress, you'll probably need a lot more.

Instructions for a vitamin C flush: Take two or three teaspoons of vitamin C powder, mix with water or fruit juice and drink. Half an hour later, repeat. If you don't have diarrhea 30 minutes after the second dose, do it again. If you haven't gotten diarrhea after three doses, begin to take 2 to 3 teaspoons of vitamin C powder every 15 minutes. If you still haven't experienced diarrhea after another hour, increase the dosage. Stop once you get diarrhea. Keep track of how much vitamin C you take. This will help you determine your optimal dosage—it's three-quarters of the amount it takes to produce a vitamin C flush—and how much vitamin C to take the next time you do a flush.

Low-Temperature Saunas and Steams

Low-temperature saunas or steambaths are useful to eliminate fat-soluble chemicals from our systems. They are commonly used to help detoxify people who have had high exposure to pesticides, solvents, pharmaceutical drugs and petrochemicals. Slow, steady, sweat encourages the release of fat-soluble toxins through the skin from their storage sites in our tissues. Most saunas and steambaths are set at temperatures too high to accomplish this, so be sure the dry sauna is set between 110 and 120°F and the steambath at 110°F so you can stay in for at least 45 minutes without getting too hot or chilled. It is best to spend 30 to 60 minutes in a sauna/steam at least three to five times a week. Releasing toxins cannot be accomplished with higher heat or shorter amounts of time. After you are done sweating, you must shower immediately afterwards using a glycerin-based soap such as Neutrogena. It will wash away the toxins and keep you from reabsorbing them.

You can perform this detoxification in a variety of ways. To use the sauna at your local health club, make sure you can use it first thing in the morning so you can control the temperature without disrupting other people who want it hotter. Or purchase a home steam cabinet. A home sauna can be made in almost any closet with professional services or kits. During the summer, you could probably accomplish the same detox by spending an hour lying in the sun or by sitting in a hot car—just not too hot! No research studies have been done on these home methods, but they would probably work if used consistently. The object is to sweat slowly and steadily.

If you have extreme toxicity to environmental chemicals, you'll need to detox under the supervision of a physician. The temporary release of toxins into your circulation can be quite severe and debilitating. Some clinics specialize in using saunas for medical detoxification. In her excellent book, *Poisoning Our Children*, Nancy Sokol Green describes her experience in a detox clinic in depth: "On the fourteenth day of detox, I started experiencing allergic symptoms, such as eyelid swelling, while I was in the sauna! . . . I was actually beginning to reek of the pesticides that had been sprayed in my home. . . . Several of the patients at the clinic who were sensitive to pesticides had to stay away from me as I triggered adverse reactions in them."[7]

As Russell Jaffe says, "There is an equation of health. In its elegant simplicity it may seem too simple. However, the validity of this equation is shown both in the practical utility of its application and in the absence of any scientific evidence against it.

Health = (Nutritional + Behavioral Competence) - (Distress + Toxic Load)."[8]

Chapter 10

Diet Means "Way of Life"

"We are what we repeatedly do. Excellent, then, is not an act, but a habit."

Aristotle

Once you have completed a cleansing diet, you are ready to begin eating anew! Your challenge is to find a way of eating and living that suits your emotional and physical needs. The word "diet" comes from Greek and means "a manner of living" or "way of life." The Latin root means "a day's journey." The key is to make real changes—changes you can live with successfully on a long-term basis—in the way you approach food, fitness and the challenges and opportunities of living.

In this chapter I will outline a plan that can help you take charge of your health and feel good on every level. Because I don't know you and your specific needs, you will have to modify these general recommendations. If you have a specific digestive condition—candidiasis, chronic fatigue, migraines, ulcers—look at the reference material for that condition and incorporate those ideas into this plan. For instance, if you have Crohn's disease, you'll probably want to avoid grain and dairy products totally. If you have arthritis, you may find that tomatoes, peppers, eggplant and potatoes aggravate your condition. Integration of your specific needs with a general life plan will give you the best results.

CLEANING OUT THE PANTRY

Go into your cabinets, refrigerator and freezer and toss out any foods that contain hydrogenated vegetable oil, vegetable shortening or partially hydrogenated vegetable oils. If you read labels, you'll find them in margarine, cookies, crackers, cereals, frozen foods, packaged foods, breads, snack foods, salad dressings, mayonnaise and so on. Because they are used so widely, your pantry may end up looking like Old Mother Hubbard's.

Why am I asking you to throw out food? Hydrogenated oils are used in foods because they are cheap, have a long shelf-life and give a buttery texture to foods. While this enables manufacturers to use inexpensive materials to reap high profits, these oils are extremely unhealthy for your body.

Liquid oils are turned into solid vegetable shortenings by the hydrogenation process. Hydrogenation converts the naturally occurring "cis" form of fat molecule into a "trans" form. When our body tries to use trans fats, they block normal biochemistry, inhibiting the function of enzymes that are involved in the synthesis of cholesterol and fatty acids. In an analogy, trans fats are like using the wrong key to open a lock and then having the key break in the lock. The trans fats jam the position so the cis fats don't fit. Trans fatty acids are found nowhere in nature and have been associated with hardening of the arteries (atherosclerosis), some types of cancer and all inflammatory illnesses—like arthritis, eczema, irritable bowel syndrome and more. Recent research indicates that these fats play at least as large a role in heart disease as do saturated fats.

Trans fatty acids also affect our body's electrical circuitry. European research has shown that essential fatty acids found in the cis formation are necessary for electrical and energy exchanges that involve sulfur-containing proteins, oxygen and light. Trans fatty acids are not suitable in these processes and jam the "plug" for the cis fats. These electrical currents are responsible for all body functions, from the way our minds work to heartbeat, cell division, muscle coordination and energy levels.

Trans fats are stickier than cis fats, says Udo Erasmus, "increasing the likelihood of a clot in a small blood vessel causing strokes, heart attacks, or circulatory occlusions in other organs, such as lungs, extremities, and sense organs."[1] Our hearts use fatty acids as their main fuel. Trans fats are less easily broken down by enzymes and have slower use as an energy source, which could have serious consequences in a high-stress situation. Trans fatty acids also interfere with our liver detoxification pathways.

Each cell in our bodies has a fatty membrane around it. Cell membranes that incorporate hydrogenated fats lose their flexibility and become more rigid, because trans fats are fairly solid at body temperature. This interferes with normal cellular function. Trans fats also change the permeability of the cell membranes, which allows substances to enter where they don't belong. If we eat hydrogenated oils, and the average American consumes 6 to 8 percent of his or her total daily calories in them, then our body fat is comprised of these fatty acids.

While you're cleaning out your kitchen, toss out the following foods: high-sugar foods, highly processed foods including white flour products, foods that contain a lot of food additives and colorings and foods that have a shelf life that extends into the next millennium. If you feel guilty about tossing these foods out, donate them to a soup kitchen. If you feel angry, write to the manufacturers and tell them to improve the quality of the foods they produce. It's a bit more expensive to eat healthful foods, but we are worth it! We can spend the money for excellent foods or we can spend it on medications and medical bills!

REPLACE YOUR PANTRY

Now you're ready to go shopping. It's important to read food labels carefully. It will take you longer the first few times you go shopping, but soon you'll be zipping through the store with your new food recognition. The foods that you bring into your home need to provide excellent nourishment. If you're careful about what you eat at home, the treats you eat at parties and restaurants will be indulgences you can feel good about. The idea isn't to be perfect, but to make progress and build good health habits slowly.

There are great treasures waiting to be uncovered in your local market and health food store that add flavor, fun and variety to your foods. If you've never been to a health food store, the salespeople will be happy to show you around and steer you to products that meet your specific needs.

RULES FOR EATING, COOKING AND SHOPPING

1. Eat local foods in season, when possible.
2. The life in foods gives us life.
3. Plan ahead. Carry food with you.
4. Eat small, frequent meals to sustain even energy levels.
5. Eat when you are hungry; stop when you are satisfied.
6. Relax while eating.
7. Choose organically grown foods whenever possible.
8. Eat as many fruits and vegetables as possible.

9. Think of protein foods as condiments rather than the main course.
10. Drink clean water.
11. Respect your own biochemical uniqueness.
12. Increase high-fiber foods.

1. Eat local foods in season. Local foods are fresh and have the highest levels of nutrients. They also have the largest energy fields and the greatest enzyme activity. Ask the produce manager at your supermarket to purchase locally grown products whenever possible. Make it part of your routine to shop at local farm stands and farmer's markets. Eating local foods in season is the concept behind macrobiotics. Because Japanese people brought macrobiotics to this country, we think of it as "eating Japanese foods," but that isn't the original intention.

Eating foods in season also helps cut down on the amount of pesticides and herbicides we consume. Foods that are flown in from outside the country—grapes from Chile, coffee from South America, bananas from Mexico—are not regulated by the same pesticide standards as foods grown in the United States. Often we get back on imported produce the very pesticides that we banned.[2] Eating local foods also helps the local economy and the environment by cutting down transportation costs and reducing consumption of fossil fuels. Think locally, act globally!

2. The life in foods gives us life. Food is fuel. Food gives us energy. Because we really *are* what we eat, if we eat foods that have little enzyme activity, they don't "spark" our body to work correctly. Enzymes are to our body what spark plugs are to the engine of our car. Without those sparks, the car doesn't run right. So if a food isn't biologically useful, who needs it? Not only do high-sugar foods not provide any nutritional benefit except calories, but sugar takes nutrients like chromium and B-complex vitamins to metabolize it. When we eat a sugar-laden diet, we continually take nutrients from our "bank account," so over time our reserves become depleted.

3. Plan ahead. Carry food with you. I've found that planning ahead and carrying food with me are two of the greatest tools for healthful eating. If you carry a bag lunch, you know it'll be healthful because you made it at home where you have only healthful foods. Just put in some leftovers or a quick sandwich with a salad and/or a piece of fruit. Add a beverage and you've got lunch. Special containers and zip-loc bags simpify the process. Planning meals ahead of time also helps you consolidate shopping trips and ultimately conserves time.

4. Eat small, frequent meals to sustain energy levels. Snacking is the best trick I know for boosting energy levels. People in Europe, South America and Japan stop in the middle of the afternoon to have tea. Our children rush home from school and eat us out of house and home. Only American adults are too busy to stop. If you find that from 3 to 6 PM it's difficult to concentrate, try this simple trick: Have something to eat in the middle of the afternoon and again just before you leave work. Here are a few quick snack ideas: Eat half a sandwich you saved from lunch plus a piece of fruit, a bagel and cream cheese with tomatoes, a rice cake with peanut butter and apples, a cup of soup and several pretzels or a handful of nuts and raisins. You'll find your energy level will stay more constant throughout the day.

5. Eat when you are hungry; stop when you are satisfied. This sounds like a simple statement, but often we eat when we aren't hungry because we're lonely, angry, depressed, bored or because we're at a social event and everyone else is eating. Before you eat anything, ask yourself the simple question: Am I hungry? If you are, then eat. If you aren't, divert your attention to other activities. Eating when you're not hungry contributes to poor digestion. Let your body use what it has before you put more into it.

6. Relax while eating. Many times we don't even stop long enough to sit down when we eat Yet eating is a time of rejuvenation of body and spirit. Take a few extra minutes to enjoy and relish the food that you eat. Take a few moments to reflect on your day and your life. Doing this can help keep your whole day in balance.

One way I've found to encourage peace of mind during meals is to say grace. It puts me in touch with the bounty of the earth we live on, makes me pay attention to the people I am with and be grateful for their presence in my life, helps me thank the people who produced the food and reminds me that we all depend upon each other and on community.

7. Choose organically grown foods whenever possible. Organic foods generally have higher levels of nutrients because organic farmers pay more attention to their animals' health and to their soils. Animals raised on organic farms are given foods that nature intended. We benefit because we don't get extra doses of hormones added to our foods. Organic farmers add more nutrients to the soil because they know that healthy plants can better fend off pests and that those nutrients end up in the foods. Bob Smith, from Doctor's Data, has released a study which analyzed organic versus commercially grown apples, pears, potatoes, wheat and wheat berries. He found that mineral levels in organically grown foods were twice as high on average as commercially grown foods.[3]

8. Eat as many fruits and vegetables as possible. The available research is overwhelming about the positive benefits of eating fruits and vegetables. They are chock full of vitamins, minerals, fiber and phytochemicals (plant chemicals) that protect us from heart disease, cancer and probably everything else. Research on phytochemicals is in its infancy, yet very promising. Citrus fruits, garlic, onions, chives, tomatoes, grapes, soybeans and other legumes, fruits, green tea, and cabbage family foods which include broccoli, kale, kohlrabi, cauliflower, collards, mustard greens, rutabaga, turnips, brussel sprouts, and bok choy—all contain phytochemicals that protect us from developing degenerative diseases. High-fiber diets also protect our colons; they reduce the incidence of polyps and bowel diseases and the risk of developing colon cancer. It's best to eat at least five half-cup servings a day, but more is better—up to 9 or 11.

9. Think of protein foods as condiments rather than as the main course. Grownups need very little protein to maintain health. It's easy to get all the protein we need from a vegetarian-based diet, so people who are eating meat two to three times a day get more protein than they need, which puts stress on their bodies. Excess protein produces uric acid and ammonia, which change our body's natural pH. This acidity affects our health and our body's ability to work optimally. And our kidneys must work harder to eliminate the urea that is produced. Studies show that vegetarians have lower incidence of heart disease, diabetes and cancer.

10. Drink clean water. Unfortunately, many cities fail to provide excellent water. The sources are often groundwater that is easily contaminated by runoff. The EPA estimates that 1.5 trillion gallons of pollutants leak into the ground each year, with the highest incidence of contaminants from lead, radon and nitrates (from fertilizers). Over 700 chemicals have been found in tap water, but testing is done on less than 200 of these.[4]

There isn't one correct answer about where to get the best drinking water. If you have a well, have the water tested for bacterial content and pollutants. Find out about your local drinking water, where it originally comes from, how it's processed and if it has fluoride added to it. Ask your water department for an analysis, and if it pleases you, drink tap water. There is much controversy today about the use of chlorine in tap water. The levels of chlorine needed to kill bacteria are rising due to increasing bacterial resistance, but chlorine has been strongly associated with elevated cancer risks. A simple water filter can efficiently remove chlorine from tap water. Activated charcoal filters are inexpensive and can remove many pollutants from tap water. Reverse osmosis is often touted as the finest system for pure drinking water, but it is more expensive than an activated charcoal filter process. Bottled water isn't necessarily any better than tap water. Often grocery store bottled water is local city tap water which has been carefully filtered with systems which are generally more efficient at removing wastes than most home filtration systems. If you buy bottled water, ask for periodic analysis of it.

11. Respect your own biochemical uniqueness. Remember the foods that are best for you are foods that agree with your body and your unique biochemistry. The rest of your family may have little or no problem eating wheat products, dairy products or any other food, but if you do, it's best to avoid them. If you really want to get well and stay well, you and your needs must come first. Your family will understand if you eat something different. If you are invited to someone's home, call several days ahead of time and let the host know you are on a restricted eating program. With advance notice, you may be able to eat nearly everything. If not, have a snack before you go and eat what you can. Perhaps you can bring a dish that suits your needs. Remember that restaurants are there to cater to you. Tell the food server if you need a menu item prepared a special way. Restaurants are becoming more used to accommodating people who have specific dietary needs.

12. Increase high-fiber foods. The connection of diet to constipation is well substantiated. Dr. Dennis Burkitt was the first researcher to connect a high-fiber diet with better health. He noticed that people eating a traditional African diet in rural areas had almost no diabetes, irritable bowel syndrome, constipation, diverticular disease, colon cancer or heart disease. In comparison, Africans consuming a Western diet had the expected incidence of these problems. In India he found a hospital where the incidence of appendicitis was only 2 percent of that in a similar American hospital. At the same hospital there was virtually no hiatal hernia, which affects nearly 30 percent of Americans over the age of 50. After looking at many factors, Dr. Burkitt concluded that the high amount of fiber in traditional diets was necessary for maintaining good health.

Known as the father of the fiber hypothesis, Dr. Burkitt made his discoveries in the 1970s. Since then, we have learned much more about fiber and how it contributes to health. For instance, we now know low-fiber diets lead to digestive disorders found in one out of four Americans. Improvement in bowel function can help prevent diverticulosis, appendicitis, colon polyps, colon cancer, hemorrhoids and varicose veins. Diets high in soluble fiber are helpful to people with irritable bowel syndrome, Crohn's disease, hiatal hernia and peptic ulcer. Dietary fiber also helps prevent obesity by slowing down digestion and the release of glucose and insulin. Fiber has been shown to normalize serum cholesterol levels. High-fiber diets reduce the risk of heart disease, high blood pressure and certain types of cancer.

Americans eat 12 grams of fiber on average a day. The National Cancer Institute recommends that we consume 20 to 30 grams of fiber daily, the same amount that Americans ate in 1850. So we are really trying to replace the fiber that was eliminated from our diet over the past 150 years! The foods that are the richest sources of fiber are whole grains (brown rice, whole wheat, bulghur, millet, buckwheat, rye, barley, spelt, oats), legumes (all beans except green beans), vegetables and fruits. These foods comprise the bulk of a healthy food plan, with nuts, seeds, protein foods and oils used as condiments. Although soluble and insoluble fibers work differently inside the body, it's important to remember that these fibers are mixed in foods. If you eat a wide variety of high-fiber foods, you will get both types of fiber without counting and measuring grams of soluble and insoluble fiber.

HEALTH BENEFITS OF DIETARY FIBER

Effect	Soluble Fiber	Insoluble Fiber
Reduce serum cholesterol	Strong	Mild
Longer full feeling	Strong	None
Sated appetite	Strong	Strong
Stabilized blood sugar	Strong	None
Reduce bacterial toxins	Strong	Strong
Speed bile acid excretion	Strong	Mild
Speed elimination	Mild	Strong
Absorb toxins	Mild	Strong
Soften stools	Mild	Strong
Improve bowel disorders	Mild	Strong
Produce gas	Strong	Mild
Block mineral absorption	None	Strong

Prepared by Martin Lee, Ph.D., *Putting the Pieces Together*, Manual for Conference, Great Smokies Diagnostic Laboratories, Asheville, N.C.

FOODS

Before I give you a shopping list, I want to discuss food basics. We eat several times each day, but few of us have studied nutrition. Most of our nutritional information has probably come from our mothers, a "health" class in high school or the media. If we believe media reports, we'll be more confused about what to eat than ever. By learning more about the foods you eat, you can begin to make a healthful food plan that will allow you to enjoy eating and feel better. Once you make a decision to rely on natural foods, your body and mind will adjust so that natural foods taste more delicious than manufactured derivatives. Once your sugar and salt taste buds calm down, fruit will taste sweet and you won't need that salt shaker as much. What

I said back in 1980 holds true today: "90% of your food should be excellent for your body, and 10% excellent for your soul."

Eating for Better Health Guidelines

These are general guidelines. If you have food intolerances, it will be important to modify these recommendations according to the needs of your body.

Whole grains	Several servings to comprise 30–45% of diet. Brown rice, oats, corn, millet, barley, buckwheat, amaranth, quinoa, wheat, teff, triticale, rye, buckwheat, spelt.
Vegetables	*Low-starch:* 3–4 cups a day. Try to include 1 cup of high calcium leafy greens such as collards, kale, dandelion, turnip greens or bok choy. Other low starch vegetables include broccoli, carrots, spinach, lettuce, onions, celery, string beans, artichoke, summer squash, endive, cucumbers, asparagus, chard, peppers, parsley, sprouts, tomatoes.
Vegetables	*High-starch:* 1 to 2 servings a day. Potatoes, yams, sweet potatoes, parsnips, winter squash, turnips.
Dried beans (Legumes)	1 or more servings a day. Split peas, lentils, kidney beans, navy beans, chickpeas, aduki beans, black beans, white beans, mung beans, soy beans, tofu.
Flesh foods	Limit to one 4–5 oz serving a day. Fish is preferable; fresh lean meats acceptable in moderation.
Dairy	0–3 servings as tolerated. Yogurt is the most easily digestible and a preferred form of dairy.
Fruits, fresh	1–3 per day (use fresh fruits in season when possible).

Essential fats 2–3 teaspoons of extra virgin olive oil or cold-pressed or expeller pressed vegetable oils of flax seed, canola, safflower, sunflower, sesame. Refrigerate all oils except olive. High temperature cooking destroys their value.

Nuts and seeds A small amount of fresh, unsalted nuts and seeds is desired. Home roasted sunflower, sesame or pumpkin seeds make an excellent snack or garnish.

Water 8 glasses a day. Hot water is best for digestion and detoxification purposes. Purified or spring water is preferred. Equally, all foods should be as toxin-free as possible. Thus locally grown, organic foods are highly recommended.

Cereals

Cereals are an excellent way to increase your fiber intake. Unfortunately, the bulk of breakfast cereals contain too much sugar, hydrogenated oils and other unhealthful ingredients, though some low-sugar, low-sodium cereals are sold in health food stores. Read package labels carefully. Check for the amount of sugar (4 grams = 1 teaspoon) and for the amount of fiber. If you have gluten or other grain sensitivities, you will be able to find many nongluten or multigrain breakfast cereals in natural foods stores. Healthy cereals are made with the whole grain, so be sure to buy those with brown rice instead of white rice or whole corn instead of degermed corn. You can also choose cereals with organically grown ingredients like oatmeal and bulghur. There are also many delicious grain blends, like 7-grain cereal, bear mush and homemade granola.

Eggs

Eggs have been given a bad rap because they've been linked to cholesterol. In fact, eggs have high amounts of phospholipids that are integral to our cell membranes. They are a precursor to acetylcholine, an important neurotransmitter.

Many researchers now believe that eating eggs has little or no effect on the serum cholesterol levels of people who have normal serum cholesterol. Recent studies found no significant change after six weeks in the cholesterol levels of the healthy people after eating two hard-boiled eggs daily.[5] Other studies have concurred that eating eggs can actually raise the "good" HDL cholesterol.[6] Current thinking is that if your cholesterol levels are normal, it's fine to eat egg yolks that have not been oxidized (exposed to oxygen)—eggs that are hard-boiled, soft-boiled or poached—since oxidized cholesterol can damage arteries and raise blood pressure. It's probably more healthful than harmful.

Studies linking high blood cholesterol levels to heart disease have proven to be only *one* accurate indicator of risk for developing heart disease. Only 25 percent of our cholesterol comes from the foods we eat. We manufacture the rest in our liver. Guided by a self-regulating mechanism, a healthy liver makes more cholesterol if we eat less and less if we eat more. From a holistic viewpoint, high serum cholesterol levels are indicative of liver distress. Eating a low-cholesterol diet may help, but if your liver is overproducing cholesterol, it usually isn't enough to correct the problem. Indeed, the link between dietary cholesterol and serum cholesterol levels has never been adequately demonstrated.

Cutting out hydrogenated oils and saturated fat can have more far-reaching effects on serum cholesterol than counting eggs. Eating fish like salmon and mackerel that are high in Omega 3 fatty acids a few times a week can lower cholesterol levels. And a whole foods diet generally contains more trace nutrients which stabilize blood fats. Learning to meditate or fully relax can also significantly lower serum cholesterol, and a regular exercise program and weight loss help normalize both LDL and HDL cholesterol.

Because egg production in this country is big business, it's best to buy eggs from naturally raised chickens at your local natural foods store, from local farmers and at farmer's markets. Commercially raised chickens live in unnatural settings, never seeing the light of day, never touching the earth beneath their cages, pumped with hormones to stimulate growth and antibiotics to prevent disease. If you believe that the quality of a seed determines the health and vibrancy of the plant it produces, then by the same token if a chicken is artificially manipulated to produce eggs, how good is the quality of those eggs? For more information on this subject, the best resource is *The New American Diet* by John Robbins.

Fish and Seafood

Fish have easily digestible protein, many trace nutrients, high quality essential fatty acids, low cholesterol levels and low saturated fat levels. This makes them a nearly ideal source of dietary protein. Fish that contain high Omega 3 fatty acids are essential to our good health. EPA (eicosapentaenoic acid) and DHA (docosahexaenoic acid), the Omega 3 fatty acids found in cold-water fish, are found in all the cells of our body. They are especially abundant in brain cells, nerve synapses, retina, adrenal glands and sex glands. If we are healthy, we can make EPA and DHA oils from Omega 3 oils found in foods. But many of us have decreased ability to make this conversion, especially people who are aging or who have a chronic illness or degenerative condition. Eating fish gives us these necessary fatty acids directly. The fish richest in this oil are salmon, halibut, tuna, mackerel, trout, sardines, eel and herring. Low-fat fish or fish from tropical waters are still healthful to eat, but they do not have any significant levels of EPA/DHA oils.

Studies show that people who eat high-fat fish twice a week have reduced risk of heart disease and stroke. Other studies have found these fish oils are protective or therapeutic for high serum cholesterol levels, angina, high blood pressure, rheumatoid arthritis, psoriasis, eczema, multiple sclerosis, asthma, lupus, allergies, inflammatory disorders, migraine headaches, and the prevention and treatment of cancer.[7] Although supplementation with fish oil capsules is beneficial, the best way to get these oils is by eating the fish itself.

It's important when selecting fish to know where the fish came from. Do not eat fish found in polluted or questionable waters. Fish and shellfish found close to shore, in rivers or lakes may be environmentally contaminated. Because the fish we eat are high on the food chain, they have the ability to concentrate environmental toxins. Mollusks that filter water, like oysters, clams, mussels and scallops, can concentrate pesticides up to 70,000 times the concentration of seawater. The FDA has set limits for pregnant women in eating certain fish, like bluefish and red snapper, because they cause birth defects. Half the world's fish catch is fed to livestock, further concentrating and passing on environmental contaminants. Farm-raised fish have fats that resemble the foods they are raised on. Because Omega 3 fatty acids tend to go rancid, other types of oils are used in fish feed, lowering the amount of beneficial oils found in these fish.

Seaweeds, also called sea vegetables, are nutritious foods that are often neglected in American cuisine, though we have been exposed to them through macrobiotics and Japanese culture. Because we evolved from the sea, sea vegetables and weeds contain nutrients that are nearly identical in ratio and quality to those we use best. Many people find that eating sea vegetables gives them a tremendous energy boost, probably because the seaweed is filling some minute nutrient need.

Meat and Poultry

Meat and poultry are nutritious and delicious, but they play far too large a role in the typical American diet. We need to think of these foods as side dishes, rather than main courses. They are generally high in fat, contain unwanted dietary cholesterol and contribute to the development of chronic degenerative diseases. It is well recognized that we are statistically healthier when limiting or excluding meats from our diets.

Environmentally, food animals are a disaster. Most water pollution comes from runoff from animal farming. The antibiotics they are given go into our rivers and streams, fields, crops and our bodies. The hormones they are fed affect us after we eat their meat. This doesn't even touch on the inhumane conditions in which these animals are raised and slaughtered.

It takes 16 pounds of grain to produce 1 pound of beef, which is an extremely wasteful use of our global natural resources. If we as a nation would consume just 10 percent less meat, there would be additional stores of grain to feed 60 million people each year.[8] Countries that switch from a grain-based to a meat-based economy become poorer, have more hunger and starvation and strip their land of natural resources.

By changing the foods you eat, you have an opportunity to help the world become a better, gentler place. For more information on this subject, read *Diet for a Small Planet* and *Food First* by Francis Moore Lappe and *May All Be Fed* and *Diet for a New America* by John Robbins.

Oils, Nuts and Seeds

When purchasing oils, it is essential to find ones that have been expeller pressed, cold pressed or are extra virgin (for olive oil). These oils have undergone a simple process of warming and pressing unlike most grocery store oils which have been chemically treated with solvents, heated to temperatures which destroy all natural antioxidants, deodorized, bleached, refined, degumed, defoamed and preserved. One should also avoid all products that contain hydrogenated oil, partially hydrogenated oil, margarine or vegetable shortening. Even margarine found at health food stores may contain hydrogenated fat.

Extra virgin olive oil is widely available in grocery stores. It retains many of the original benefits of the olives themselves: betasitosterols, which balance serum cholesterol, and beta-carotene, components that help the liver, gallbladder and pancreas, brain and blood. Olive oil is mostly monounsaturated. Both mono- and poly-unsaturated fats help prevent heart disease, but monounsaturated fats don't have the damaging effects of large amounts of polyunsaturated fats.

Other oils are found at health food stores and some gourmet food shops. Keep all oils except olive oil in the refrigerator so they don't go rancid. Good quality oils vary in flavor and color—walnut oil tastes different from corn oil. You'll find it's best to use less of the stronger flavored oils.

Though the fat in nuts and seeds provides minerals and essential fats, use them sparingly. Too much can harm you and put on unwanted pounds. A cup of nuts contains between 800 and 1200 calories, most of which come from fat. Chestnuts, which are delicious steamed or roasted, are an exception to this rule. A teaspoon as a garnish on salads or vegetables can add a lot of flavor and nutrients without adding too much fat. Buy raw or roasted nuts that have not been seasoned, dry roasted or coated with sugar and/or salt. These are available in the baking section of most stores and in health food stores. Store nuts and seeds in the refrigerator or freezer to keep them fresh.

Peanuts are in the bean family and grow underground rather than on trees. Some peanuts have a mold, called aflatoxin, which is toxic and can cause cancer. Aflatoxin is tasteless, so it's impossible to tell if it's on your peanuts or peanut butter. A healthy person should be able to handle any aflatoxin they're exposed to, but a sick person may have a difficult time. Be sure to buy only the old fashioned kind—ground up peanuts and salt. Almond butter, cashew butter and sesame tahini are good alternatives.

Shopping List

Here is a shopping list of healthful foods. It includes some brand names I know, but hundreds more are equally fine. Because "natural" and organic foods are often less available and sometimes more expensive than other foods, many people have banded together to form informal food co-ops. Together, they can order from wholesale distributors of natural foods, and buy foods in bulk or by the case. Buying foods in bulk saves time and money, and most grocery and health food stores are happy to give a discount when you buy cases.

Healthy Food Choices

Fruits & Vegetables

Eat at least 5 servings a day of all fruits and vegetables. They are rich in nutrients and fiber, contain no cholesterol and are low in fat. Olives and avocados are rich in essential fatty acids (oils), so if you're watching your weight, go easy!

Legumes (Beans & Peas)

Use all types freely and often.

Black beans
Great northern beans
Split peas, green & yellow
Lima beans
Fava beans
Kidney beans
Pinto beans
Lentils: green, yellow, orange
White beans
Garbanzo (chickpeas)
Soybeans: tofu, tempeh, miso, flour, grits

WHOLE GRAINS

Many people with digestive illness have difficulty handling grains of any type. If so, you may need to eliminate this food group from your diet for a period of time while you heal.

*Denotes a grain that contains gluten.

- Brown rice
- Whole wheat:*
  - whole wheat flour,* wheat berries,* cracked wheat,* bulghur,* wheat bran,* couscous*
- Wild rice
- Buckwheat
- Barley*
- Rye*
- Oats*: oatmeal,*
  - whole oats,* oat bran*
- Millet
- Spelt*
- Quinoa
- Kamut
- Corn: flour,
  - meal, bran, polenta, pop corn
- Amaranth

WHOLE GRAIN BREADS

Found in some grocery stores, bakeries and natural foods stores. Look for "whole wheat" or "stone ground wheat" on the label. If it says "wheat flour" or "enriched" flour, it's white flour. Dense, heavy breads usually contain the most nutrients and fiber.

GLUTEN-FREE BREADS & PASTAS

Many stores are now carrying breads and pastas made with gluten-free flours like quinoa, rice, millet and soy. These products are terrific for people with celiac disease or wheat sensitivities.

Breads: whole grain
Pasta: whole wheat, corn, quinoa, semolina, Jerusalem artichoke, rice, soba, udon, cellophane noodles
Crackers: Akmak, Health Valley, Hain, Finn crisp, plain Rye Krisp, Barbara's pretzels & breadsticks, Wasa (rye only), whole wheat matzoh
Tortillas, corn/wheat, English muffins, pita bread, rice cakes, whole grain muffins, pancakes, waffles

CEREALS

Health Valley
Barbara's Bakery
Kashi
Perky's
Nature's Path
Shredded wheat with bran
New Morning
Uncle Sam's cereal
McCann's Oatmeal
Nutrigrain
All Bran
Erewhon
Oatmeal
7-grain cereal

DAIRY PRODUCTS & EGGS

Many people with digestive problems need to eliminate dairy products from their diet, so use only if you are able to digest them easily. Use low-fat products as a rule. Cheese is best used as a condiment. Try some grated parmesan or other types of cheese to zip up a dish, rather than eating hunks of cheese as snack food.

Low- or non-fat yogurt
Low-fat cottage cheese
Goat's milk and cheese
Eggs: chicken, duck
Skim milk
Buttermilk
Parmesan
Butter & ghee in small quantities
Skim milk ricotta
Kefir
Yogurt cheese

FISH & SEAFOOD

When possible, choose fish that are highest in Omega 3 fatty acids. All other fish and shellfish are good low-fat protein choices, but will not give the protective benefits of high Omega 3 fish.

- Salmon: all types
- Tuna: albacore, bluefin
- Herring
- Lake trout
- All other fish & shellfish
- Mackerel
- Atlantic sturgeon
- Anchovy
- Eel
- Seaweed & sea vegetables
- Bluefish (sparingly)
- Sablefish
- Sardines

MEATS & POULTRY

Choose low-fat cuts of meat and take the skin off poultry.

- Lean hamburger & beef
- Pork tenderloin
- Chicken, without skin
- Ground turkey
- Venison
- Turkey, light meat, no skin
- Leg of lamb
- Buffalo

BEVERAGES

Make certain your drinking water is uncontaminated. Fresh juices made from organic produce offer a delicious and healthful option, and buy organic juices and teas whenever possible. Check labels carefully for sugars added to juice drinks; only 100 percent pure juices can be labeled "juice."

Pure water	Grain coffee	Fruit juices
Vegetable juices	Pure carbonated waters	Green tea
	Herbal teas	

Oils, Nuts & Seeds

Use only extra virgin or expeller pressed oils, organic oils whenever possible and olive oil, except for stir frying or baking.

Oils

Olive oil	Sesame oil	Canola oil
Sunflower oil	Safflower oil	Corn oil
Walnut oil	Almond oil	Soybean oil
Peanut oil	Rice bran oil	Macadamia nut oil

Nuts & Seeds: Use sparingly.

Pistachio	Macadamia	Cashews
Almond	Chestnut	Pecans
Walnuts	Peanuts	Sunflower seeds
Pumpkin seeds	Sesame seeds	Pine nuts
Coconut, unsweetened		

Condiments

Condiments add great flavor to foods, with few calories or grams of fat. Read labels carefully; many sauces are salty, fatty, sugar laden or filled with food chemicals. Homemade salad dressings are best.

Mustard	Horseradish	Salsa
Worcestershire sauce	Tamari or shoyu sauce	Vinegars: balsalmic, rice, fruit

Herbs & spices
Anchovy paste
Sun dried tomatoes
Herbal blends in shakers
Pepper: white or black
Pickles
Gomasio
Capers
Olives: black & green
Seaweed flakes
Sauerkraut

Chapter 11

Skills for Developing Emotional Hardiness: Stress Management

"Things which matter most must never be at the mercy of things which matter least."

GOETHE

RECENT DISCOVERIES HAVE proven beyond a doubt that our brains and immune systems are interlinked. Our thoughts affect us on a cellular level. The field of psychoneuroimmunology is a blossoming discipline which studies the mind-body connection. More fatal heart attacks occur on Monday than any other day of the week. Could this have something to do with our attitude about Mondays? Researchers believe it does.

We experience stress on three levels: physical, environmental and mental/emotional. In preceding chapters we've talked a lot about physical and environmental stress, so here we will focus on the mental and emotional aspects. All our lives have stressful periods. We can't eliminate them, but we can learn to flow with them more easily. It is often the meaning we attach to our experiences that makes them stressful or stress-free.

Stress plays a major role in many digestive problems. Anyone who has ever had diarrhea from "nerves" can attest to that. The butterflies that people feel in their stomach when they're nervous happen to all of us. Continued stress in our body, mind or meanings affects the body's ability to heal and perform. If the body is using most of its energy to put out fires, it has less time for maintenance and repair. Because the digestive tract repairs and replaces itself every few days, it is one of the first places where our bodies alert us that all is not well. People who have digestive-related problems can benefit from stress management tools.

Though we mainly think of stress as a negative thing, it does have a positive side. The stress of going to college, getting married, having children or getting a promotion provides opportunities and challenges that force us to change, grow and strive to fulfill our potential.

Stress = Opportunity

But even positive stress can overwhelm us, causing distress. When we bite off more than we can chew, we feel stressed out. In many instances we have little control over stressors like illness, loss of a job, financial worries and death. But even in these circumstances we can have control over our thoughts and behavior. This is where stress management offers many benefits. Studies have shown that people who have emotional hardiness handle distress more easily than people with less resilience. Hardy people take on a challenge, make a commitment, take control and see what happens. If success comes their way, it encourages them to try new things. If they fail, they pick themselves up and try again, recognizing that experience gained from "failure" adds to life's perspective. Fortunately, it's possible for anyone to learn hardiness skills.

Mental stress is one of the greatest challenges to our immune systems, putting pressure on nearly every organ and system in the body. When we feel stressed out, our bodies react with an increased heartbeat, shallow and rapid breathing, a release of adrenaline, raised blood sugar levels and oxygen rates. Our muscles tense so we can move quickly. An increased blood supply is sent to our brain and major muscles, with decreased blood flow to our extremities. Even our pupils dilate and we sweat more. Our bodies react this way because our minds tell us there is a dangerous situation that requires quick thinking and movement. Historically, this might have been a bear on our path, a forest fire or the exhilaration of the hunt. Today, it can be anything from a near car collision, three phone lines ringing at once to burnt toast! Because our first reaction is to unconsciously hold our breath and breathe shallowly, deep breathing exercises are an excellent stress management tool. Breathing deeply brings more oxygen to our tissues while waste products are excreted. It slows us down, so we feel more balanced and centered and can make clearer decisions. Deep abdominal breathing is something you can do anywhere, and no one can tell!

Abdominal Breathing Exercise

Take ten deep breaths through your nose. As you inhale, feel your belly expand; feel the inhalation spread until it fills your lungs to the top. Exhale by letting your lungs deflate gently. Remember that there aren't any awards for how big your lungs get or how many seconds you can hold your breath! Notice how you feel. If you're calmer, remember this tool and use it as needed.

How do you know when you're under too much stress? Your body will usually tell you before your mind does. You may get a neckache, backache or headache, a sick feeling in your stomach, hives, fatigue or a myriad of other symptoms. But even though your body is telling you to stop, you keep on going. Our culture rewards this type of behavior. But we owe it to ourselves to listen to our own needs more carefully and to respond to them in kind.

SIGNALS OF STRESS

Emotional	Physical	Behavioral
Moodiness	Muscle tension	Overuse of drugs/alcohol
Worry	Aches & pains	Overeating
Irritability	Fatigue	Forgetfulness
Hostility/anger	Sleep disturbances	Clumsiness
Bad dreams	Diarrhea	Depression
Lassitude	Digestive symptoms	Tension
Defensiveness	Headaches	Poor eating
Difficulty concentrating		

We can tune into these signals, or we can choose to ignore the messages and carry on with our activities. If we ignore them, the symptoms may go away, or they may blast louder and louder until we are forced to pay attention or our bodies break down. Being attentive to small signals allows us to gently get back on track without experiencing major upheavals in our lives.

The most important component of a good stress management program is to have a plan with reasonable and realistic goals. If you need to, get a buddy or professional who will support you. Think of the ways in which you might slip or lose your way and prepare for them. Once your goals and support system are in place, you can forge into action.

SELF-RENEWAL

The best stress management tool is exercise. Exercise makes us stronger and more flexible and increases our sense of balance, which helps keep us injury free. Regular exercise helps control blood sugar levels so our energy is more sustained, and being fit lowers the risk of heart disease. While exercise alone won't make you thin, a consistent exercise program can help you move toward your ideal body weight. In addition, our bodies release endorphins, morphine-like molecules, in the brain which make us feel happy! When I ask people to describe the benefits of their exercise program, they tell me they have more energy, feel higher self-esteem and are more relaxed.

Positive thinking is an important part of stress management. If you could tape the conversation in your brain for an hour or two, you might find you had a lot of self-criticism or self-doubts. With practice we can easily learn to "flip" these negative images and turn criticism into a positive thought or plan of action. When we catch ourselves playing a negative tape, we need to eject it and put in a new tape. Instead of thinking "My ulcerative colitis will get worse and worse until I need surgery," you can flip the image and say, "So far I haven't licked this problem, but if I am persistent, I can improve my health."

Play the Hit Side of the Tape Exercise

Close your eyes for a minute, and think of all your most wonderful attributes. Compliment yourself freely. Take some time to appreciate your good points and achievements. Think about times in your life when you helped others, fell in love and really felt good about yourself. Make a mental list of the nicest things you can say about yourself. It's okay if you repeat yourself.

After you finish, see how you feel, and repeat this exercise as often as you like. If you want, write your ideas down, and put them somewhere so you'll be reminded of how terrific you are! Liking ourselves also helps our view of others and the world around us.

Most of us invest a lot of energy in our work, home, family

and friends. But stress management means you need to make time to be nice to yourself. Take an hour every day to recharge your batteries. It's not important what you do. Each of us will find renewal in different things. Here are some ideas: read something for fun, play a musical instrument, listen to music, garden, exercise, be outdoors, have a date with a friend, write a letter, keep a journal, enjoy a hobby, take a class, go to church/temple, meditate. Vacations are an important way to put our lives in perspective, to value what is truly important. Learn to nurture yourself.

Prioritizing helps us find the balance point in our lives. Balance is hard to achieve and maintain, but it is an honorable goal. Like many people, if something really interests me, I take on new responsibilities and enjoyable events until I become overwhelmed. Then I make a list of all of my commitments and prioritize them to see what I can let go of responsibly. Soon my life is back into balance—until the next exciting possibility comes along and I'm overcommitted again. Be assertive: Learn to say no!

On the other hand, many people never grab for the ring and watch the world go by because they're worried about taking a risk. For these people, it's important to find a spark of interest and act on it. Doing nothing because we're fearful of failure can be as stressful as overachieving.

The demands placed upon us in this fast-paced world are unrealistically high. We are expected to play many roles perfectly—mate, son/daughter, mother/father, businessperson, athlete, spiritual being and public citizen, both locally and globally. After all, the Greek and Roman gods and goddesses were excellent at only one thing. So why do we put such unreasonable and unrealistic demands on ourselves? A Buddhist saying is: "Expectation is the root of all suffering." If we can be easier on ourselves and in our relationships, we can find more love, contentment and peace.

Compartmentalizing thoughts is a useful tool. Since we can only think about or act upon one responsibility at a time, it helps to put each one in a "compartment." The pearl here is to be able to be 100 percent present and focused on each task while you are doing it. This frees the mind and calms the spirit.

Compartmentalization Exercise

Imagine you are sitting before a rolltop desk with pigeonholes in it. In each hole is a scroll tied neatly with a ribbon. See yourself opening one scroll at a time, examining it, putting it back in its proper place, selecting another and repeating the process. Do this for several minutes. Find a way to use these pigeonholes in your daily life to help you accomplish tasks step by step. Practice this exercise to help you create closure between tasks and events in your life.

Quick Ideas for Stress Management

Eat healthy foods.
Develop better communication skills; learn to really listen.
Exercise regularly.
Spend time outdoors.
Make time for yourself each day for pleasure or relaxation.
Meditate or learn self-hypnosis or visualization techniques.
Realize that you don't have to be perfect.
Think creatively.
Go at your own pace.
Think of solutions, not problems.
Prioritize.
Keep journals.
Live one day at a time.
Play and laugh.
Spend time with friends and family.
Be flexible
View your problems as an opportunity for growth.
Plan for chaos in your daily schedule.

Set clear priorities and stick to them.
Breathe deeply.
Plan ahead: wake up 15 minutes earlier each morning, keep your car in good working order, put a duplicate car key in your wallet and so on.
Learn to say "No!"
Unplug your phone or let your answering machine pick up the calls.
Take a bath, shower, steam, or Jacuzzi.
Make lists. Keep a calendar or appointment book.
Drive 10 mph slower.
Take one day a week to relax. Honor the Sabbath.
Believe that people have good intentions and are doing the best job they can.
See the world through a "loving" filter.

Part 5

A Functional Approach to Therapeutics/ Self Care

Chapter 12

Natural Therapies for Common Digestive Problems

"Disease bias means that we take health for granted, waiting to act when health is gone and disease emerges. Once we make this assumption, we can soon become so preoccupied that our horizon is filled with diseases to combat. Because disease looms so large, our sight is obscured to the possibilities of health."

Russell Jaffe, M.D.

This chapter provides a comprehensive list of self-care ideas for the most common digestive problems. The remedies are mostly nutritional and herbal because those are the fields I know best; I have included other modalities whenever possible. The most important ones are listed first. Read each section that applies to you, find the remedies for your symptoms, look especially for repetitive ones and try those first. Also try the remedies that make the most sense intuitively.

Each herb or nutrient is listed separately, but often they can be found in combination supplements. You'll notice that specific recommendations are repeated for many problems. Although each health condition has its own unique properties, many have similar characteristics that respond to similar treatment programs. You may want to work with a health professional to tailor a program that will best suit your needs.

Health care is both a science and an art. You may need the science in the form of lab-testing, diagnosis and evaluation of your needs. (Your doctor will order the customary lab work. I have included information about functional lab tests which are most likely to reveal new information; these tests will probably be unfamiliar to your physician. The Directory will help you connect with the appropriate labs.) The art of healing comes into play when determining which paths to follow, which ideas have most merit and which dosages are appropriate. If the first program you try doesn't work or only works partially, try another. You can feel better if you are persistent and patient.

This chapter contains classic digestive problems. We start our journey at the mouth and move south. Some of the following ideas alleviate symptoms, while others work to help your body heal the underlying cause. You can begin your program by taking a multivitamin and mineral supplement. Be sure to purchase one that is hypoallergenic.

Multivitamin with minerals: Think of a multivitamin with minerals as inexpensive health insurance, and arm yourself with an excellent supplement. Your diet is likely to be deficient in several nutrients which it can provide. Because minerals are bulky, you'll find yourself taking anywhere from four to nine pills daily. Look for a supplement which contains the following: 1000 mg calcium, 500 mg magnesium, no more than 400 IU vitamin D, a minimum of 100 IU of vitamin E, a minimum of 250 mg of vitamin C, 100-200 mcg chromium, 100-200 mcg selenium, 5-10 mg manganese, at least 15 mg zinc, and at least 25 mg of each B-vitamin.

Mouth

The mouth is the first digestive organ (see Chapter 3). The health of our teeth, tongue and gums is integral to the health of the rest of the digestive tract. Digestive enzymes in saliva begin the process of carbohydrate digestion, and chewing sends signals to the brain, which in turn sends signals to the stomach that food is on the way. Thorough chewing of food can help with indigestion.

Irritation and inflammation in the mouth can be signs of food or chemical sensitivities or allergies. The mouth is our first contact with ingested allergens. Careful investigation of the mouth area can give information about a person's nutritional status. Cracks down the center of the tongue are an indication of the need for increased B-complex vitamins. Bleeding gums indicate the need for vitamin C and bioflavonoids. Receding gums indicate bone loss, so bone nutrients are needed.

Bad Breath/Halitosis

Yes, bad breath can be digestive in origin. It can be caused by low HCl levels in the stomach, poor flora and/or constipation. But, first, consult a dentist to see if it is caused by poor dental hygiene, periodontal disease or tooth infections. If so, follow your dentist's advice, and also look in the section on gum and tooth health which follows. If your gums and teeth are healthy, look to your digestive capacities. Using mouthwash is like putting a band-aid on a broken leg.

Healing Options

Eliminate constipation: See section on stool transit time in Chapter 3 and do the self-test. Make sure you are getting plenty of fiber and liquids and are having one to two bowel movements each day. Magnesium is essential for normal peristalsis. Dosage: 400-800 mg magnesium daily.

Probiotic supplement: Take 1-2 capsules of acidophillus and bifidus between meals.

If the problem persists: This may be a sign that you are fermenting rather than digesting your foods. (1) To check out your HCl levels, try 1 tsp vinegar in a glass of water with meals or betaine HCl tablets. If the HCl causes burning, you probably don't need HCl. You can neutralize the acid with milk or baking soda. (2) Ask your doctor to run a comprehensive digestive and stool analysis with parasitology evaluation. You may have dysbiosis, parasites or a Helicobacter infection (the bacteria implicated in ulcers). A CDSA can help you find out what's amiss.

Cracks in the Corners of the Mouth and Lips

Our skin is continuously replacing itself, and the places where our skin folds need to be replaced even more often. B-complex vitamins, particularly vitamins B2 (riboflavin) and B6 (pyridoxine), assist in formation of new skin. Cracks at the corners of our lips, called cheilosis, are most often associated with these nutrient deficiencies. They can easily become infected by yeast (*Candida albicans*). If they do not respond to nutritional therapy, have a physician look for other causes.

Healing Options

B-complex vitamins: Dosage: 50-100 mg one to three times daily in trial for four weeks.

Gingivitis and Periodontal Disease

Gingivitis is an inflammation of the gums which if left alone often progresses to periodontitis, an inflammation of the bone around the teeth. Periodontal disease increases with plaque build-up, age, long-term use of steroid medications, and in diabetics, people with systemic disease and smokers. The presence of silver fillings, which contain 50 percent mercury, has also been found to predispose people to periodontal disease. One study showed that when silver fillings were removed, 86 percent of the 125 oral cavity symptoms were eliminated or improved.[1]

Gingivitis and periodontitis are complex problems which have complex solutions. Periodontal disease will affect 9 out of 10 Americans during their lifetimes, and 4 out of 10 will lose all their teeth. Regular dental care is essential. Follow your dentist's advice and practice consistent oral hygiene: brush and floss daily.

Nutrition plays a critical role in dental health. One recent study looked at gingivitis, plaque adhesion and calculus deposit with regard to the eating habits of teenagers. They concluded that teenagers with diets adequate in nutrients had better oral health than teenagers with diets that contained fewer nutrients.[2]

Teeth are made of bone material and need the same nutrients for rebuilding as other bones. It has long been considered that receding and inflamed gums were a sign that people brushed too hard, causing damage to the gums, but new theories propose that gums recede because bone throughout the body, including the teeth, is demineralizing. If other bones need 17 nutrients to remineralize, the same goes for teeth. Calcium alone cannot reverse the problem.

Vitamin C deficiency causes bleeding gums and loose teeth and contributes to gingivitis. One symptom of scurvy, a vitamin C deficiency disease, is bleeding gums. We rarely see outright scurvy in our population, but we often see people with bleeding gums. Vitamin C is also important for bone formation and collagen synthesis and is essential for gum repair. Vitamin A is also necessary for collagen synthesis and formation of gum tissue.

Other researchers look to zinc deficiency or a low zinc to copper ratio as the culprit in gum disease. Zinc is integral to maintenance and repair of gum tissue, inhibits plaque formation and reduces inflammation by inhibiting mast cell release of histamine. It also plays a role in immune function.

Vitamin E has been used clinically for periodontal disease. Bacterial plaque, long known to be a culprit in tooth decay and gingivitis, produces compounds which weaken and irritate the gum tissue. They include endotoxins and exotoxins, free radicals, connective tissue-destroying enzymes, white blood cell poisons, antigens, waste products and toxins.

Antioxidant nutrients and co-enzyme Q10 have been associated with improved gum health, reduced periodontal pocket depth and less tooth movement. Bioflavonoids make the tissues stronger, reduce inflammation and cross-link with collagen fibers, making them stronger. Because bioflavonoids work synergistically with vitamin C, bleeding gums often respond to vitamin C and bioflavonoid supplementation.

Folic acid, a B-complex vitamin, is important for maintenance and repair of mucous membranes. The need for extra folic acid was first noted for pregnant women, while subsequent studies have shown that it plays an important role for gingival health in all people.

Healing Options

Improved diet: Focus on fresh fruits, vegetables, whole grains and beans. Foods rich in flavonoids are beneficial: blueberries, blackberries, purple grapes.

Multivitamin with minerals: Since you are depleted in many nutrients, arm yourself with an excellent multivitamin with minerals. Because minerals are bulky, you'll probably take anywhere from four to nine pills daily. Look for a supplement which contains the following: 1000 mg calcium, 500 mg magnesium, no more than 400 IU vitamin D, 100-200 mcg chromium, 100-200 mcg selenium, 5-10 mg manganese, at least 15 mg zinc, and at least 25 mg of each B-vitamin.

Co-Enzyme Q10: Dosage: 75 mg daily for a trial period of three months.

Antioxidants: Vitamins C and E, selenium, glutathione, super oxide dismutase (SOD), beta-carotene and other antioxidant nutrients are depleted in diseased gum tissues. Supplementation can facilitate repair. For ease of use, purchase an antioxidant supplement which contains many antioxidants. Dosage: Use as directed for three months.

Vitamin C: Dosage: 500-1000 mg one to three times daily.

Bioflavonoids: Use quercitin, bilberry (blueberry), grapeseed extract or pycnogenol for their antioxidant and anti-inflammatory effects.

Myrrh: Myrrh has been used since biblical times. It has soothing and antiseptic properties for mucous membranes.

Mouth Ulcers/Canker Sores

Mouth sores are common. Most of us have experienced mouth ulcers, canker sores or cold sores, but some people have chronic problems with them. Usually found inside the mouth, canker sores, called *aphthous stomatitis*, or aphthous ulcers, are the result of poor intestinal flora, food sensitivities/allergies, stress, hormonal changes and nutritional deficiencies. High-sugar foods and high-acid foods, like pineapples, citrus and tomatoes, sometimes trigger canker sores.

If you have reoccurring canker sores, thoroughly investigate the possibility of food sensitivities. Also, make sure your toothpaste, mouthwash and floss aren't causing the problem. A study showed that use of Piroxicam, an NSAID, caused mouth ulcers which resolved when the patient was taken off the medication.[3] If you have canker sores that don't resolve after several weeks, let your doctor or dentist examine you.

Healing Options

Allergies and sensitivities: Cigarettes, toothpaste, mouthwash and flavored dental floss can cause irritation. Make sure they are not the source of your problem. Food sensitivities often are. Rule them out carefully with the elimination/provocation diet and/or food allergy/sensitivity blood testing.

Probiotics: *Lactobacillus acidophillus* is often beneficial in prevention and treatment of canker sores. Dosage: 1-2 capsules or 1/4-1/2 tsp of the powder three times daily; take between meals.

B-complex vitamins: Deficiencies in vitamin B1, B2, B6, B12 and folic acid have been associated with recurrent canker sores. People with B-complex deficiencies showed significant improvement of mouth ulcers during three months of supplementation with B-complex vitamins.[4]

Gluten sensitivity: Gluten is a protein fraction found in wheat, rye, spelt, oats and barley. A considerable amount of research has been done on the connection between gluten intolerance and mouth ulcers because people with celiac disease (sprue) often have recurring mouth sores. About 25 percent of people with chronic canker sores have elevated antibodies to gluten, which indicates a specific sensitivity. When they avoid gluten-containing grains, their mouth sores go away.[5]

Iron deficiency anemia: Iron deficiency is associated with canker sores. If you get recurrent canker sores and are anemic, you may respond to iron supplementation. People who are not anemic will not benefit. Ask your physician to test you for anemia. Dosage: 30-75 mg elemental iron daily. Because iron tends to be constipating, a slow-release iron, like Feosol or generic equivalents, may be helpful. Floradix, an herbal iron supplement, is gentle and works well. Cooking in cast iron pots and pans is another way to gain iron from your diet; it also provides other trace nutrients.

Stress management: Ask yourself if stress plays a significant role in your canker sores. If so, consider yourself lucky and use the information discussed in Chapter 11 to make changes in your lifestyle that will benefit all of you, not just your mouth!

Zinc: Zinc deficiencies have been linked to mouth ulcers. Zinc plays an important role in healing wounds and immune system function. In one study zinc supplementation helped heal canker sores 81 percent of the time in people with low zinc levels or a low zinc to copper ratio.[6]

Topical Remedies

Ice: Ice compresses dry up canker sores quickly. Apply ice directly to the sore for either 45 minutes once a day or several times a day for five minutes each. You'll still have a scab that needs to heal, but the sores won't be painful.

Licorice root: Licorice root is soothing to the mucous membranes of the digestive tract, and chewable licorice can help reduce inflammation and pain from mouth ulcers. Licorice promotes healing of mucus membranes by stimulating production of healing prostaglandins. Just be sure to buy DGL licorice which has had the glycyrrhizins removed. Glycyrrhizins can raise blood pressure and lower serum potassium levels. Chew 2 licorice tablets between meals as needed up to four times daily.

Myrrh: Myrrh is an herb that has been used since biblical times to soothe mucous membranes. It has antiseptic properties and can be used in a variety of ways. Chewing gum with myrrh can be temporarily soothing, and a glycerin tincture can be used topically to soothe the sores. It can be combined with the herb golden seal in tea, paste or tincture.

Golden seal: Golden seal is soothing to mucous membranes and also has antiseptic properties. It can be taken internally, dabbed directly on the sores or drunk as a tea.

Castor oil: An old Edgar Cayce remedy is to soak a cotton swab in castor oil and apply to the canker sore.

Thrush

Thrush is a yeast infection in the mouth and throat. It has a white, cottage cheesy look and is common after use of antibiotics. Thrush can be treated with either prescription or natural medicines. If it persists, you must treat yourself systemically. It is of primary importance to use probiotic supplements of acidophillus and bifidus to reestablish normal mouth/throat flora. Natural remedies such as garlic, grapefruit seed extract, pau d'arco and mathake tea, along with dietary changes, can make your body inhospitable to candida. Follow the protocols for candida infections. In one study, one-third of people with thrush were found to have folic acid, vitamin B6 or vitamin B12 anemias,[7] so it's worth having your doctor check you for anemias.

Tongue Problems

Glossitis is an inflammation of the tongue, which can be extremely red and smooth. People may also develop other symptoms; the most common is called "geographic tongue" where the center of your tongue looks like a miniature Grand Canyon. Look in a mirror, look at your friend's and family's tongues, and you'll probably find one. People may also have scalloping on the edges of their tongues.

Tongue problems can arise from systemic illness, so celiac, diabetes, anemia and syphilis should be ruled out by your physician. More often tongue problems are indicators of nutritional needs or mouth irritants, like smoking or other chemicals. Studies have found that glossitis is a sign of protein-calorie malnourishment, nutritional deficiencies or marginal nutritional deficiencies of several vitamins and minerals.[8] It most often signals the need for increased B-complex vitamins and iron. You will often find a reddened tongue with pellagra, which is caused by a deficiency of niacin (vitamin B3). Therapeutics for glossitis are almost identical to those for canker sores.

Healing Options

B-complex vitamins: The most important B vitamins for tongue health are riboflavin (B2), niacin (B3), vitamin B12 and folic acid. Choline is found in B-complex vitamins and also plays a vital role in tongue health. Dosage: 50-100 mg one to three times daily for a trial period of four to six weeks.

Iron deficiency anemia: This can also cause a sore and inflamed tongue. Have your physician check to make sure your iron status is normal.

Zinc: Zinc is important for healing. Dosage: 25-50 mg daily.

ESOPHAGUS

The most common problems in the esophageal area are belching, medically called "eructation," and heartburn, also called gastric reflux.

Belching/Eructation

Belching is a symptom of gas in the upper part of the digestive tract. It is a release of trapped air from the stomach and usually comes from swallowed air. Just as a baby needs to be burped if he/she swallows air, we also burp if we swallow air—it's normal. Other than being culturally embarrassing, it's usually without problem. In fact, in China it's considered polite to belch after a meal—it means you really enjoyed it!

Foods and drink that contain air contribute to belching. Without fail if I have a carbonated drink, I burp. Whipped cream and egg whites have the same effect. Gulping drinks and foods causes us to take in more air, while eating slowly prevents us from swallowing air. People also swallow air during exercise and while chewing gum, sucking on pipes or cigarettes. If you are overweight, you are more likely to belch from exercise.

Be thankful that you belch; air trapped in the stomach can be painful and belching is a safety valve which relieves the pressure. If you have a problem with the amount of belching you do, here are some suggestions.

Healing Options

Eat slowly. chew your food well.

Avoid carbonated beverages.

Stop smoking: If you smoke, stop. Be glad you have such a benign reason to stop.

Reach and maintain ideal body weight: If you are significantly overweight, lose some weight.

Stop chewing gum or sucking on candy.

Heartburn/Gastric Reflux and Hiatal Hernia

Heartburn is caused by stomach acid backing up into your esophagus. The esophageal sphincter is supposed to keep the stomach contents in place, but if the sphincter relaxes, acid can push up into the esophagus. The most common symptoms are a burning sensation above the stomach, excessive salivation, belching, regurgitation and a sour taste in the mouth. One third of the people in America experience frequent heartburn, and 3 to 7 percent suffer from esophageal illness caused from the acid reflux which causes scarring, constriction of the esophagus and swallowing disorders. Some drugs can cause heartburn, including birth control pills, diazepam, nicotine, nitroglycerine, progesterone, provera and theophylline.

Four to five million Americans seek medical advice each year for heartburn and hiatal hernia. Heartburn is common among pregnant women whose organs are squashed in a most peculiar way. For most people, heartburn is a mild, self-limiting problem, yet for 20 percent of the people affected, it becomes a serious health problem. Stress plays some part in it. Other things that trigger heartburn include wearing tight fitting clothes, lying down, bending over and eating large meals or specific foods. If you experience heartburn in the middle of the night, be sure to eat at least four hours before going to bed.

Hiatal hernia occurs when a portion of the stomach gets pushed through the diaphragm and into the thoracic cavity where it doesn't belong. Hiatal hernias may or may not cause symptoms, the most common of which is heartburn. It's found in about 20 percent of all middle-aged Americans. Dr. Dennis Burkitt hypothesized that hiatal hernia was a contemporary problem and the result of a modernized diet. Gastric reflux and hiatal hernia have been on the rise over the past 20 years and occur most often in people of Caucasian descent. These problems are rarely seen in people eating high-fiber diets. Dr. Burkitt felt that the pressure of straining with bowel movements pushed the upper part of the stomach out of place. Chiropractic adjustment can gently put the stomach back in place, and in many cases only a single adjustment is necessary.

People commonly take antacids for temporary relief of heartburn. Initially, use of antacids causes the body to produce more hydrochloric acid that helps digest food. Parietal cells respond by making more acid. Eventually the parietal cells get exhausted so, over the long term, antacids cause the parietel cells to make less hydrochloric acid. Hans Selye, father of modern theories about stress, devised the model that before an organ breaks down, it overworks. There are other repercussions of taking antacids. Most bacteria can't live in a high-acid environment and are killed in the stomach. Low stomach acid predisposes people to dysbiosis. Antacids also decrease your stomach's ability to digest protein by reducing the effectiveness of protease enzymes.

Healing Options

Osteopathic care and chiropractic adjustment: Seek chiropractic care for hiatal hernia. Cranial-sacral adjustments can often correct gastric reflux, especially in children. Chiropractic or osteopathic adjustment is often all the therapy you need for these problems.

Dietary: Eat healthy foods. Increase fruits, vegetables, grains, beans and high-fiber foods. Drinks and foods that are more acidic, like tomatoes and citrus, are more likely to cause heartburn. Dairy products have been shown to trigger symptoms, but more symptoms were provoked with milk with higher fat contents, suggesting that fat was the culprit, rather than the milk. Alcoholic beverages, coffee and, to a lesser extent, tea provoked heartburn, as well as high-fat, fried and spicy foods and chocolate. Trigger foods are individual—you need to discover what yours are.

If you are overweight, lose weight.

Place a 6" beam under the head of your bed: If you suffer from night-time heartburn, raising the head of your bed can alleviate symptoms. Although you might think that raising your bed would feel strange, the difference is barely noticeable, and the heartburn improves.

Possible *Helicobacter pylori* infection: *Helicobacter pylori* is a bacteria which has been implicated in gastric and duodenal ulcers. In some cases it is also involved in gastric reflux. Treatment with antibiotics and bismuth containing supplements or drugs can cure *H. pylori*.

Hydrochloric acid: Heartburn has traditionally been treated with antacid therapy, but often it responds well to supplementation with hydrochloric acid pills. Often the symptoms of excess stomach acid and decreased stomach acid are the same. To test yourself, dilute a tablespoon of vinegar with water and drink with meals. If the vinegar is not strong enough to show results, take 1 to 2 HCl tablets with meals. If you do not need the HCl, you will feel a burning sensation which can be neutralized with milk or baking soda. If you need it, you'll feel relief of symptoms.

Cabbage juice: Cabbage juice has been a long-standing folk remedy for heartburn. Its high glutamine content is probably the key to its success. Cabbage juice has a strong flavor, so dilute with other vegetable juices.

Glutamine: Although I was unable to find any references for use of glutamine for heartburn, it makes theoretical sense. The digestive tract uses glutamine as a fuel source and for healing. It is effective for healing stomach ulcers, irritable bowel syndrome and ulcerative bowel diseases, and it makes sense that it would be useful in the upper GI as well. Dosage: Begin with 8 g daily.

Slippery elm bark: Slippery elm bark has demulcent properties, and it's gentle and soothing to mucous membranes. It has been a folk remedy for both heartburn and ulcers in European and Native American cultures and was used as a food by Native Americans. It can be used in large amounts without harm. Drink as a tea, or chew on the bark. Recipe for tea: Take 1 tsp of slippery elm bark in two cups of water. Simmer for 20 minutes and strain. Sweeten if you want, and drink freely. You can also purchase slippery elm lozenges at health food stores and some drug stores.

Lobelia: Massage tincture of lobelia externally onto the painful area and take 2 to 3 drops internally. This is a remedy recommended by Dr. Christopher, one of the greatest American herbalists of our times.

Ginger: This herb can provide temporary relief in a tea. Steep 1/2 tsp of powdered ginger or 1 1/2 tsp of fresh ginger per cup of boiled water for 10 minutes and drink. If you like, sweeten it with honey. Use freely.

Meadowsweet herb: Also a demulcent, meadowsweet soothes inflamed mucous membranes. To make a tea, steep 1 to 2 tsp of the dried herb in one cup of boiled water for 10 minutes. Sweeten with honey if you like. Drink 3 cups daily.

STOMACH

Gastric Hypofunction/Hypochlorhydria

Hypochlorhydria has been associated with many common health problems. Not all of them are considered "digestive" problems, but they are. As we age the parietal cells in the stomach lining produce less hydrochloric acid. In fact, half of people over the age of 60 have hypochlorhydria (low stomach acid), and by age 85, 80 percent of the healthy people tested had low stomach acid.[9]

The balance of stomach acidity is vital to the health of the stomach and to our overall health. Hydrochloric acid (HCl) is our body's first line of defense against disease-causing microbes. Low HCl levels open us to the possibility of food poisoning and dysbiosis along the digestive tract. Overgrowth of bacteria in the intestinal tract occurs in 20 percent of people aged 60 to 80 and in 40 percent of people over age 80.[10] Adequate HCl is necessary for the absorption of vitamin B12 from food. B12 deficiency causes weakness, fatigue and nervous system problems. Several minerals require an acidic environment for absorption, including iron, calcium, magnesium, zinc and copper, and most B-complex vitamins require normal levels of stomach acid. Vitamin C levels are also low in people with poor stomach acid. Acid is critical for the breakdown of protein bonds in the stomach. Poor acid content in the stomach causes indigestion. The symptoms of hypoacidity often mimic those of hyperacidity.

FUNCTIONAL LABORATORY TESTING

1. Heidleburg Capsule Test: This test measures your ability to produce hydrochloric acid when challenged with alkaline substances. See Chapter 7 for complete details.

Low Gastric Acidity

Common symptoms of low gastric acidity

Bloating, belching, burning and flatulence immediately after meals
A sense of fullness after eating
Indigestion, diarrhea or constipation
Multiple food allergies
Nausea after taking supplements
Itching around the rectum
Weak, peeling and cracked fingernails
Dilated blood vessels in the cheeks and nose (in nonalcoholics)
Acne
Iron deficiency
Chronic intestinal parasites or abnormal flora
Undigested food in stool
Chronic candida infections
Upper digestive tract gasiness

Diseases associated with low gastric acidity

Addison's disease	Lupus erythematosis
Asthma	Myasthenia gravis
Celiac disease	Osteoporosis
Dermatitis herpetiformis (herpes)	Pernicious anemia
Diabetes	Psoriasis
Eczema	Rheumatoid arthritis
Gall bladder disease	Rosacea
Graves disease	Sjögren's syndrome
Chronic autoimmune disorders	Thyrotoxicosis
Hepatitis	Hyper- and hypothyroidism
Chronic hives	Vitiligo

Healing Options

Betaine HCl: Dosage: Begin with one 10 mg capsule of betaine HCl with meals. If you do not respond, build slowly to a maximum of 5 capsules with each meal. If you experience burning, immediately neutralize the acid with 1 tsp baking soda in water or milk. That indicates that you now have too much HCl and are irritating your stomach lining. Cut back your dosage to a comfortable level.

Vinegar: You can increase stomach acidity with vinegar. Dilute 1 teaspoon of vinegar in water and drink with each meal. Gradually increase the amount of vinegar, up to 10 teaspoons. If you experience burning, immediately neutralize the acid by drinking a glass of milk or taking a teaspoon of baking soda in water.

Vitamin B12 deficiency: Have your physician test your serum B12 levels. If your levels are in the high-normal range, you don't have a deficiency. If your levels are low-normal, you may still be deficient in vitamin B12. More sensitive tests of vitamin B12 status are methylmalonic acid and homocysteine levels, but these tests are not widely available and are fairly expensive. A more expedient and less expensive route is to ask your physician to give you 1000 mcg of vitamin B12 by injection weekly for four weeks. Then you and your doctor can evaluate the benefits. You may use sublingual vitamin B12 lozenges to prolong the effects of the vitamin B12 shots.

Multivitamin with minerals: Adequate HCl is necessary for absorption of vitamins and minerals. Since you are depleted in many nutrients, arm yourself with an excellent multivitamin with minerals. Because minerals are bulky, you'll probably find yourself taking anywhere from four to nine pills daily. Look for a supplement which contains the following: 1000 mg calcium, 500 mg magnesium, no more than 400 IU vitamin D, 100-200 mcg chromium, 100-200 mcg selenium, 5-10 mg manganese, at least 15 mg zinc, and at least 25 mg of each B-vitamin.

Digestive enzymes: I recommend plant-derived enzymes because they are able to work in the low pH of the stomach and in the neutral environment of the intestines. They provide protease and lipase for the stomach and serve your enzyme needs throughout the digestive tract. Dosage: 1 to 2 with meals for a trial period of four weeks.

Swedish bitters: Bitters are a long-standing remedy for poor digestion in Europe. They stimulate production of hydrochloric acid. Take bitters either in tablet or liquid form as needed.

Chew food thoroughly.

Eat small meals frequently: Small meals are easier to digest.

Avoid drinking liquids with meals: Fluids dilute stomach acid.

Gastric Ulcers and Gastritis

A 1994 National Institutes of Health statement reports that "Peptic ulcer disease is a chronic inflammatory condition of the stomach and duodenum that affects as many as 10 percent of people in the United States at some time in their lives. The disease has relatively low mortality, but it results in substantial human suffering and high economic costs."

Gastric ulcers occur in the stomach and the duodenum (the first section of the small intestine) where gastric juice has burned a hole in your lining. It hurts! Gastric juice, which has a 0.5 percent concentration of hydrochloric acid, would burn your hand if you spilled some on it. A mucus layer protects the stomach tissue from being eaten away by the protein-splitting enzyme, pepsin, and gastric juices. Secretions of bicarbonate (baking soda) from the stomach lining are mixed into the mucus, buffering the acid. This makes an effective barrier to keep the stomach lining from harm. Pepsin, the real villain in this story, slowly digests this mucus layer, and if the mucus isn't replaced, gastric juices come into contact with the stomach lining and ulcers occur.

Until recently it was believed that ulcers were the result of high stomach acidity and high stress levels. Therapy consisted of stress management and a bland diet. In the 1970s receptor sites on the stomach lining were found which regulate secretion of hydrochloric acid, and drugs were developed to block these receptor sites. These drugs, called H2 blockers, have been effective in healing ulcers but not in preventing reoccurrence.

In 1982, Australian physician Barry Marshall discovered the presence of a bacteria, *Helicobacter pylori* (*H. pylori*), between the stomach lining and the mucous membrane. This bacteria is found in nearly all people with duodenal and stomach ulcers and with gastritis.[11] *H. pylori* is also found in people who do not have any evidence of ulcers or gastritis. Are these people eventually going to develop problems? Could these people have a different strain of *H. pylori* which isn't pathogenic? We don't know; stay tuned for the next decade to find out.

What is known is that when *H. pylori* is eradicated, ulcers heal and don't reoccur. The treatment now used is a combination of antibiotics, bismuth and anti-parasitics, and ulcers are healed 90-97 percent over a one-year period. Bismuth is an elemental mineral which protects the stomach lining by protecting the mucous membrane from being dissolved by pepsin. Although this treatment has minor side effects, the overall outcome shows improved quality of living and less psychological stress after therapy.[12]

Because ulcers have been experienced throughout history, people have found effective natural therapies. Most physicians are not aware of these therapies, but nutritionally oriented physicians have been using them with promising results. For example, Leo Galland, MD, has been using a combination of antibiotic therapy and bismuth, with DGL licorice, citrus seed extract, golden seal, activated charcoal and aloe vera.[13]

Functional Laboratory Testing

1. *Helicobacter pylori* test.

HEALING OPTIONS

Licorice: DGL (deglycyrrhized) licorice helps heal the stomach mucus lining by increasing healing prostaglandins that promote mucus secretion and cell proliferation. Licorice enhances the blood flow and health of intestinal tract cells. It's important to use DGL licorice to avoid side effects caused by whole licorice.

Grapefruit or citrus seed extract: Citrus seed extract has widely effective antiparasitic, antiviral and antibiotic properties.

Golden seal: Golden seal is soothing to mucous membranes, enhances immune function and has antibiotic and antifungal properties.

Aloe vera: Aloe vera is a folk remedy for ulcers and has been approved by the FDA for use in oral ulcers. It is soothing and healing to mucous membranes.

Gamma-oryzanol: Gamma-oryzanol, a compound found in rice bran oil, is a useful therapeutic tool in gastritis, ulcers and irritable bowel syndrome. It acts on the autonomic nervous system to normalize production of gastric juice and has also been shown to be effective in normalizing serum triglycerides and cholesterol, symptoms of menopause and depressive disorders.[14] Studies involving 375 hospitals in Japan indicate that gamma-oryzanol was effective in reducing symptoms from 80 to 90 percent, with over half of the participants experiencing total or marked improvement. Typical dosage was 100 mg three times daily for three weeks. Occasionally the dosage was doubled, and often the therapy was used longer. Minimal side effects were experienced by 0.4 percent of the people.[15]

Rice bran oil is widely used in Asia as a cooking and salad oil and is available in the United States at some gourmet stores. Dosage: 100 mg of gamma-oryzanol three times daily for a trial period of three to six weeks.

Cabbage juice: Cabbage juice is a long-standing folk remedy for heartburn. Dosage: Drink 1 quart of cabbage juice daily for a trial period of two weeks.

Glutamine: Glutamine is the most popular anti-ulcer drug in Asia today. The digestive tract uses glutamine as a fuel source and for healing. It is effective for healing stomach ulcers, irritable bowel syndrome and ulcerative bowel diseases. Dosage: Begin with 8 g daily for a trial period of 4 weeks.

Slippery elm bark: Slippery elm bark has demulcent properties and has been a folk remedy for both heartburn and ulcers. It can be used in large amounts without harm. Drink as a tea, chew on the bark or take in capsules. Recipe for tea: Simmer 1 tsp of slippery elm bark in two cups of water for 20 minutes and strain. Sweeten if you want, and drink freely. Dosage: 2 to 4 capsules three times daily for a trial period of three weeks.

Sano-Gastril: Sano-Gastril is a chewable tablet which buffers the acidity of the stomach. (Sano-Gastril is marketed in this country by Nutri-Cology/Allergy Research Group; this is not an endorsement, but the only product of its type.) Sano-Gastril is composed of an extract called glycine-max and a specific strain of *Lactobacillus acidophillus* plus vitamin C and other nutrients. A study using two tablets three times daily was done on 93 people with ulcers and gastritis. After one month each participant was X-rayed to see progress. At that time 12 out of 22 people with gastric ulcer, 25 out of 58 people with duodenal ulcer and 4 out of 12 people with gastritis were completely healed.[16] Two tablets of Sano-Gastril three times daily before meals relieved heartburn completely within 5 to 10 minutes in 76 percent of 158 people.

Evening primrose oil: Evening primrose oil increases the levels of prostaglandin E2 series, which promote healing and repair. Dosage:1000-2000 mg three times a day for a trial period of four weeks.

Flax: Low dietary intake of the essential fatty acid linoleic acid has been associated with duodenal ulcers. Flax seeds and oil are excellent sources of linoleic acid. A benefit to using ground flax seeds rather than the oil is that the mucous portion of the flaxseed buffers excess acid, which makes it ideal for inflammation in the stomach and throughout the gastrointestinal tract. Linoleic acid is also found in pumpkin seeds, tofu, walnuts, safflower oil, sunflower seeds and oil, and sesame seeds and oil. Dosage: 2 to 3 tsp Flax oil goes rancid in a matter of weeks. Store it in the freezer to keep it fresh.

Zinc: Zinc increases the rate of healing and can prevent damage to the stomach lining. Dosage: 50-100 mg daily.

Vitamin A: Vitamin A is protective and promotes healing of gastric ulcers. Dosage: 10,000-25,000 IU daily. Pregnant women should not exceed 10,000 IU vitamin A.

Pancreas

The pancreas secretes bicarbonates into the duodenum. This is essentially baking soda which neutralizes the acidity of chyme so that it won't burn the internal tissues of the intestines. The pancreas also secretes enzymes which digest carbohydrates, protein and fats. Proteins are broken down by these molecules into single amino acids. When protease enzymes are doing their job, they also protect us from allergic reactions caused by protein particles being absorbed into the bloodstream which trigger an antigenic response. Poor production of pancreatic juice is called pancreatic insufficiency.

The signs of pancreatic insufficiency include gas, indigestion, bloating, discomfort, undigested food in our stools, undigested fat in our stools and food sensitivities. It is common in people with candidiasis or parasite infections and is an underlying cause of hypoglycemia. Pancreatic insufficiency also increases as we age. People with pancreatitis and cystic fibrosis have pancreatic insufficiency. Stool testing by complete digestive stool analysis provides an indirect measure of pancreatic function by measuring chymotrypsin, a pancreatic enzyme for digestion of protein, and by measuring how well you are able to digest meats and vegetables. The Chymex test is also a measure of pancreatic activity.

Causes of pancreatic insufficiency are stress (mental and physical), nutritional deficiencies, poor diet, eating only cooked foods, exposure to radiation or toxins, hereditary weaknesses, drugs and infections.

FUNCTIONAL LABORATORY TESTING

1. Comprehensive digestive stool analysis (CDSA).
2. Chymex testing for pancreatic function.
3. Food allergies—IgE.
4. Elisa testing for food sensitivities—IgG.

HEALING OPTIONS

Eat in a relaxed manner.

Chew your food thoroughly.

Limit beverage intake with meals: Drinking liquids at meals dilutes the gastric juices in the stomach and pancreatic juice in the small intestines.

Pancreatic enzyme supplements: Clinical experience shows that pancreatic enzymes work well as a digestive aid. Glandular-based supplements, like pancreatic enzyme preparations, are directed to specific tissues, helping to initiate repair. Pancreatic enzymes also help restore the balance of GI flora. In studies done on monkeys, it was shown that pancreatic enzymes were able to kill *Clostridium*, *Bacteroides*, *Pseudomonaceae*, *Enterobacter*, *E. coli*, and *Klebsiella*.[17] Continued use of pancreatic enzymes can help with repair and maintenance of pancreatic tissue.

Pancreatic tissue from pigs has been widely used over the past 50 years to supply missing pancreatic enzymes. You can purchase products from health food stores or ask your physician to prescribe them. The United States Pharmacopoeia (USP) regulates the strength of pancreatic enzymes. For best results use a preparation that is 4x to 10x. Dosage: 1-2 tablets/capsules at the beginning of meals.

Vegetable enzymes: For people who would rather have a vegetarian alternative to pancreatic enzymes, vegetable enzymes are a suitable option. These enzymes are derived from a fungus called *Aspergillus oryzae* which is used to ferment soy sauce, tamari and miso.

These enzymes work in a much wider range of pH than pancreatic enzymes, enhancing digestion in the stomach as well as in the intestines. Because they are not broken down by stomach acid, the required dosage is much smaller than that for pancreatic enzymes. Vegetable enzymes are also less likely to cause food allergy reactions. Some companies include probiotic bacteria in daily supplements. Dosage: 1-2 capsules at the beginning of meals.

GALLBLADDER

The gallbladder, a pear-shaped organ that lies just below the liver holds bile for the liver, which produces it. The gallbladder's function is to store and concentrate bile which emulsifies fats, cholesterol and fat-soluble vitamins by breaking them into tiny globules. These create a greater surface area for the fat-splitting enzymes (lipase) to act on for digestion. Between meals the gallbladder concentrates bile.

Gallbladder problems usually indicate poor liver function. The liver is the most complex organ in the body. Unlike the heart which has one major function—to beat—the liver has a multitude of functions which include regulation of blood sugar levels, making 13,000 different enzymes, "humanizing" food by acting as a filter, breaking down toxins, manufacturing cholesterol and bile, breaking down hormones and more. Because of its complexity and the 10,000 pounds of toxins it must filter over a lifetime, the liver can easily become overwhelmed.

When we eat a meal that contains fat, the liver and gallbladder are stimulated to release bile, which contains bicarbonates (baking soda) that alkalize stomach acid in the duodenum. Each day the liver secretes about a quart of bile, which is absorbed into the body from the ileum and colon and returned to the liver to be used again. Drugs and other toxins are eliminated from the liver through bile. The brown color of stool comes from the yellow color of bilirubin in bile.

Gallstones

The most common digestive problem associated with the gallbladder is gallstones. One in five Americans over the age of 65 has gallstones, and half a million surgeries are performed each year for removal of the gallbladder. Women are two to four times more likely to be affected by gallstones than men. However, most people who have gallstones are never bothered by them, and doctors only treat them if they are causing problems. An inflammation of the gallbladder can lead to pain and discomfort. Symptoms can range from abdominal discomfort, bloating, belching and food intolerances. When you have more than one stone, you may experience a sharp pain or a spasm under the ribs on the right side. Occasionally the pain will be felt under your right shoulder blade. These pains are often strongest after eating a high-fat meal.

Diet plays an important role in prevention of gallbladder disease. Low-fat, low-meat and vegetarian diets are recommended, as is a low-sugar, high-fiber intake. Dennis Burkitt, a British M.D. who lived and worked in Africa for 20 years, found that he performed only two surgeries for gallbladder removal among Africans eating an indigenous diet.[18]

If you are overweight, losing weight will reduce your risk of developing gallstones. Just be careful because several studies have shown that fasting and *extremely* low-fat, low-calorie diets increase your risk of developing gallstones.[19] Fasting for more than 14 hours raises the risk of problems due to gallstones. So easy does it while dieting; always eat breakfast.

Medical treatment for gallstones consists of an injection of a drug that dissolves the gallstones, oral medication to dissolve stones, lithotripshy which breaks stones with sound waves or surgical removal of the gallbladder.

Physicians familiar with natural therapies have favorable results treating gallstones without use of drugs or surgery by having their patients detoxify the liver and strengthen liver function. Metabolic cleansing or other detoxification programs are a critical first step in treatment. Food sensitivities also play an important role in the development of gallstones—most patients with gallbladder disease have them[20] and they must be identified.

Functional Laboratory Testing

1. Liver function profile.
2. Home test for bowel transit time.
3. Elisa testing for food sensitivities and IgE testing for food allergies.
4. Heidleburg capsule test for levels of hydrochloric acid production.

Healing Options

Low-fat diet: Low-fat diets help prevent gallstones and also reduce pain and inflammation associated with gallstones. Saturated fats found in dairy products, meats, coconut oil, palm oil, hydrogenated oils and vegetable shortening stimulate concentration of bile. While a low-fat diet is optimal, essential fatty acids are vital to gallbladder function and overall health. Make sure you get 1-2 tbsp of uncooked expeller pressed oils or extra virgin olive oil each day—the easiest way is in salad dressing.

Low-sugar diet: Several studies have indicated that people who consume a lot of sweets are more likely to develop gallstones.[21]

Decrease coffee intake: Coffee may trigger gallbladder attacks in susceptible people. Use of either regular or decaffeinated coffee raised levels of cholecystokinen and caused gallbladder contractions. Stop drinking coffee and see what effect this produces. Some people get horrible headaches or flu-like symptoms when they withdraw from caffeine. If you do, wean yourself gradually.

Vegetarian diets: Vegetarian diets have been found to be helpful in reducing the incidence of gallbladder disease.

Eat breakfast: Fasting for more than 14 hours raises your risk of problems due to gallstones.

Reduce bowel transit time: People with gallstones have significantly slower transit times than healthy people. Eat more high-fiber foods and drink more fluids.

Drink 6 to 8 glasses of water every day!

Rule out food sensitivities: In 1968 Breneman found that food sensitivities play a role in gallbladder disease. He put 69 patients on an elimination diet consisting of beef, rye, soy, rice, cherry, peach, apricot, beet and spinach. After three to five days all people were symptom-free. With a slow reintroduction of foods they were sensitive to, symptoms returned. The most common food offenders were eggs (3 percent), pork (64 percent) and onions (52 percent).[22] What's interesting about this study is that beef and soy are often trigger foods for food sensitivity reactions.

Rule out deficient levels of hydrochloric acid: In a *Lancet* study about half of the people with gallstones had insufficient levels of hydrochloric acid.[23] A Heidleburg capsule test can determine if you have sufficient levels of hydrochloric acid. You can also take HCl supplements with meals. Begin with one 500 mg capsule of HCl with 150 mg of pepsin. If you don't need the HCl, you'll experience a burning or warm sensation. If this is uncomfortable, you can drink some milk or a teaspoon of baking soda in water to neutralize the acid. Many people with low-acid secretion need several HCl tablets per meal; increase the dosage slowly to find the correct one for you. Too much will cause a warmth or burning sensation.

Milk thistle/silymarin (Silybum marianum): Extracts of the herb milk thistle have been used historically since the fifteenth century for ailments of the liver and gallbladder. It helps normalize liver function, detoxify the liver, which it does gently and thoroughly, and improves the solubility of bile. Silymarin promotes the flow of bile and helps tone the spleen, gallbladder and liver. Dosage: Three to six 175 mg capsules daily of standardized 80 percent milk thistle extract with water before meals.

Lipotrophic supplement: Lipotrophic supplements contain substances which help normalize liver and gallbladder functions. They may contain dandelion root, milk thistle, lecithin or phosphatidyl choline, methionine, choline, inositol, vitamin C, black radish, beet greens, artichoke leaves (*Cynara sclymus*), turmeric, boldo (*Peumus boldo*), fringe tree (*Chioanthus virginicus*), greater celandine and ox bile. Lipotrophics may also contain magnesium and B-complex vitamins (B6, B12 and folate) to enhance their function. Dosage: Use as directed on the label. Use 1000 mg each of methionine and choline daily.

Lecithin: Phosphatidyl choline, the most biologically active form of lecithin, and lecithin have been shown to make cholesterol more soluble, which reduces formation of gallstones.

Studies have shown that as little as 100 mg of lecithin three times daily will increase lecithin concentration in bile.[24] Dosage: 500 mg daily.

Vitamin C: Vitamin C has been shown to prevent formation of gallstones. People with high risk to developing gallstones have low ascorbic acid levels.[25] Dosage: 1-3 grams daily.

Liver/gallbladder flush: Anecdotal stories about people showing up at their doctor's office with a jar full of stones after a gallbladder flush are abundant, but there is little documentation to validate whether what they passed are really gallstones. Do this procedure at home *only* under the supervision of a physician.

From Monday through Saturday, drink as much natural apple juice as possible. Continue to eat normally and take your usual medications or supplements. On Saturday eat a normal lunch at noon. Three hours later (3 PM) dissolve 1 Tbsp of Epson salts (magnesium sulphate) in 1/4 cup of warm water and drink it. This is a laxative and helps peristalsis move the stones through your digestive system. It doesn't taste great so you may want to follow it with some orange or grapefruit juice. Two hours later (5 PM) repeat the Epsom salts and orange or grapefruit juice. For dinner eat citrus fruits or drink citrus juices. At bedtime, drink 1/2 cup of warm extra virgin olive oil blended with 1/2 cup of lemon juice. Go to bed immediately and lie on your right side with your knees pulled up close to your chest for half an hour. On Sunday morning take 1 tablespoon of Epsom salts in 1/4 cup of warm water an hour before breakfast. If you have gallstones, you will find dark green to light green stones in your bowel movement on Sunday morning. They are irregular in shape and size, varying from small like kiwi seeds to large like cherry pits. If you have chronic gallbladder problems, you may want to repeat this again in two weeks. This can be repeated every three to six months if you continue to form stones.

Small Intestine

The most common symptoms of small intestinal problems are gas and bloating. When you are having digestive discomfort, it is important to rule out candidiasis, food sensitivities, pancreatic insufficiency, parasites and small bowel bacterial overgrowth, which may be underlying causes of gas and bloating. Use of digestive supplements and beneficial flora can also be helpful.

Flatulence/Intestinal Gas

Everyone has gas. It's normal. In fact, we "pass gas" an average of ten to fifteen times a day. Most of our gas comes from swallowed air. Chewing gum, drinking carbonated drinks and whipped foods like egg whites and whipped cream all contribute to swallowed air. The gas we pass is mainly nitrogen (up to 90 percent), carbon dioxide and oxygen, which are odorless. Gas and bloating are also a product of the fermentation of small pieces of undigested foods by the bacteria in our intestines. Fermentation produces stinky gases like methane and hydrogen sulfide, which has the odor of rotten eggs. Other substances, like butyric acid, cadaverine and putrescine are present in tiny amounts, but they are noted for the mighty fragrance they give to gas.

Some of us experience excessive amounts of gas, which can be not only embarrassing but also an uncomfortable sign that something is out of balance. Millions of people have bloating and discomfort associated with gas. If you've ever made wine, you'll recall putting a balloon on the top during the fermentation process that allowed for expansion of the gasses produced. Our bellies act like a balloon, expanding to contain the gas produced by fermentation.

Foods from the cabbage family, dried and sulfured fruits and beans all contain sulfur which gives gas a rotten egg odor. Cucumbers, celery, apples, carrots, onions and garlic are all commonly known to cause gas. People with lactose intolerance experience gas when they eat dairy products. Eating a high fiber diet is healthful, but your intestinal flora need to adjust gradually to these dietary changes. You may have insufficient levels of hydrochloric acid, intestinal flora, pancreatic enzymes or a dysbiosis which is causing your problems. Food sensitivities, especially to wheat and grains, can also cause excessive gas.

Functional Laboratory Testing

1. Intestinal permeability test.
2. Complete digestive and stool analysis with parasitology.
3. Elisa testing for food sensitivities.
4. Lactose breath test.

Healing Options

High-fiber diet: Most of us need to dramatically increase the amount of dietary fiber we eat, but raising these levels too quickly can cause a lot of gas and discomfort. Our flora go wild with sudden increases in dietary fiber, and the fermentation of these fibers causes gas. Increasing your fiber intake more slowly will solve this problem. High-fiber foods include whole grains, beans and many fruits and vegetables.

Cooking beans: Beans are an excellent source of vegetarian protein, containing both soluble and insoluble fibers, and sitosterols that help normalize cholesterol levels. However, beans are notorious for their gas-producing effects. They contain substances which are difficult for us to digest. For instance, beans, grains and seeds hold their nutrients with phytic acid. Soaking or sprouting releases the nutrients so that we can absorb more of them. First, soak the beans for 4 to 12 hours, then drain off the water, replace with new water and simmer for several hours until they are soft. Some people find that putting a pinch or two of baking soda in the water helps reduce gas. Others add kombu, a Japanese sea vegetable, or ginger. Beano is an enzyme product that contains the enzymes necessary for digestion of beans. Place a drop or two on your food; it really helps reduce flatulence. Beano is sold widely in drug and health food stores.

We produce digestive enzymes for foods we commonly eat. If you eat beans rarely, start by eating a tablespoon or two of beans each day. Your body may begin to produce the enzymes necessary for their digestion.

Chew your food well and eat slowly: These simple activities can have far reaching effects on healthy digestive processes and gas reduction.

Lactose intolerance: The inability to digest lactose, the sugar in milk, is a frequent cause of gas. Eliminate all dairy products for at least two weeks and see if there is improvement. Make sure to eliminate all hidden dairy products found in foods. These are listed in the section on lactose intolerance in Chapter 7.

Food sensitivities: Although lactose intolerance is the most common food sensitivity, people can be sensitive to nearly any other food. The most likely culprits are sugars and grains. Careful charting of your foods and flatulence levels can help you detect which foods are giving you the most trouble.

Food sensitivities don't usually exist by themselves. If you have a number of food sensitivities, check for candida infection and dysbiosis.

Fermentation dysbiosis: An imbalance of intestinal flora often causes excessive gas. Candida fungi cause fermentation of sugars, fruits and starches which we feel as gas and bloating. A comprehensive digestive and stool analysis can determine whether or not you have a candida infection or other dysbiotic imbalance.

Digestive insufficiencies/intestinal flora and enzyme supplementation: Many people find that supplementation with digestive enzymes at meals, either vegetable or pancreatic enzymes, plus probiotic supplements containing acidophillus and bifidobacteria, really help prevent gas. Dosage: 1-2 digestive enzymes with meals.

Acidophilus and bifidobacteria: Use of beneficial bacteria supplementally can make a tremendous difference in your ability to digest foods. Beneficial flora can help reestablish the normal microbial balance in your intestinal tract. Dosage: 1-2 capsules or 1/4-1/2 tsp powder two to three times daily on an empty stomach. Mix powdered supplement with a cool or cold beverage; hot drinks kill the flora.

Sorbitol and Xylitol: Sorbitol and xylitol are undigestible sugars found in most sugarless candy and gum. They are used by diabetics and dieters because these sugars are sweet but don't affect blood sugar levels. Large amounts of sorbitol and xylitol cause gas, but even small amounts can cause a problem in people who are sensitive.

Parasites: Parasites often cause gas and bloating. If you have explored more obvious causes, you may want to have a stool test for parasites.

Chlorophyll: Chlorophyll liquid or tablets can help prevent gas. Take two to three times daily with meals.

Ginger, fennel and anise: Most of us have at least one of these spices in our kitchen, and they are valuable tools for reducing gas. Put a few slices of fresh ginger or 1/2 tsp of dried ginger in a cup of boiling water and steep until cool enough to drink. It will soon begin to dispel your gas from both ends, and you'll be much more comfortable. Fennel and anise can be used in tea or simply chew on the seeds to relieve gas. In Indian restaurants you find small bowls of these seeds. They also cleanse the palate with their sweet pungency.

Herbs and herbal teas: Traditionally herbs and spices were added to foods to aid digestion. Nearly all our common kitchen herbs and spices have a beneficial effect, including basil, oregano, marjoram, parsley, thyme, celery seed, peppermint, spearmint, bayberries, caraway seed, cardamom seed, catnip, cloves, coriander, lemon balm and sarsaparilla. You can find many digestive herbal tea blends in health food stores.

Parasites

Some symptoms of parasites appear to be like any other digestive problem. Chronic diarrhea is often a sign of a parasitic infection. Other symptoms include pain, constipation, bloating, gas, unexplained weight loss, fatigue, unexplained fever, coughing, itching, rashes, bloody stools, abdominal cramping, joint and muscle aches, irritable bowel syndrome, anemia, allergy, granulomas, nervousness, teeth grinding, chronic fatigue, poor immune response and sleep disturbances. These symptoms can come and go due to the life cycles of the specific parasite involved.

If your anus itches at night, you may have pinworms. They are most commonly found in children and parents of small children. It's fairly simple to determine whether or not you have pinworms: When you feel the itching, put a piece of cellophane tape on your anus, and examine the tape for small white wiggley worms that look like white pieces of thread that move! Or you can simply put your finger into your anus and see if you pull out any worms. With children, you can use tape or just look.

Recently there have been references in the scientific literature which suggest that parasites may be the primary cause of allergies. Parasites cause damage to the lining of the digestive tract which allows large molecules to enter the bloodstream. This provokes an antigenic response. This theory is revolutionary and additional research needs to be done to determine just how large a role parasites play in allergies.[26]

Parasite Testing

Many physicians request parasitology testing on random stool samples, but this type of testing is not very accurate.

Even with repeated samples accurate results are found only 85 to 90 percent of the time. Many parasites, like Giardia, live farther up the digestive tract so that many labs now give an oral laxative to induce diarrhea when testing for parasites. This type of sample is called a stool purge.

Other parasites are found by using a rectal swab rather than a stool sample. Because some parasites generally live in the mucous membranes that line the intestinal tract rather than in the feces, a rectal swab enables a physician to collect a sample of the mucus. Stool purge tests and rectal swabs used with random stool samples give a much more accurate picture of colon health than single or multiple random stool samples. Many labs collect a random stool sample, a purge stool sample or two and a rectal swab. The most accurate stool testing is usually done by labs that specialize in parasitology testing. Due to the high volume of samples, high-tech microscopes and slow pace, their staff have become experts in detection and recognition of parasites.

Healing Options

Prescription medication may be the most efficient treatment for most parasite infections. Within a week or two you are parasite-free. However, these medications can be hard on the liver and disruptive to the intestinal flora. After using them, it's wise to take acidophilus and bifidobacteria to replace and rebalance the intestinal flora.

Natural options work more slowly—about a month—but are highly effective. They generally contain garlic, wormwood (artemesia), golden seal, black walnut and/or grapefruit seed extract. Acidophillus and bifidobacteria should be replaced after therapy is finished.

Garlic: Historically, garlic has been used for pinworms; it has antiviral, antibacterial and antifungal properties. Allicin, the active component in garlic, has been shown to be effective against *Entamoeba histolytica* and *Giardia lambia*. *Enta-*

moeba histolytica is an amoeba which has been used to evaluate the value of anti-amoeba drugs.

Golden seal: Golden seal has been used historically to balance infections of mucous membranes throughout the body. It is also effective with candida infections. Berberine sulfate, an active ingredient in golden seal, has been shown to be effective against amoebas and giardia parasites.

Artemesia annua/wormwood: Wormwood has been used for centuries in China and Europe for worms and parasites. It contains sesquiterpene lactone which works like peroxide. It is believed to affect the parasite membranes, weakening them so our natural defense system kicks in. Artemesia also contains an ingredient which is effective against malaria even when it is resistant to quinine drugs. Tea of wormwood has been successfully used for pinworms and roundworms by Dr. Christopher, one of America's foremost herbalists.[27] Caution: Artemesia is safe when used in a tea or capsule, but pure wormwood oil is poisonous. Dosage: 1/4-1/2 tsp powdered wormwood once or twice daily or make a tea using 2 tsp of fresh leaves or tops in 1 cup water. Drink 1/2 cup a day, one teaspoon at a time.

Black walnut: The juice of unripe, green hulls of black walnuts has been traditionally used for treatment of parasites and fungal infections. Black walnut is a folk remedy for ringworm, athlete's foot and healing cracks in the palms and feet.

Dichroea febrifuga (Saxifragaceae): Dichroea is a Chinese herb called Changshan which is effective against malaria, amoebas and giardia.

Jerusalem oak/american wormseed/chenopodium anthelminticum: A folk medicine used throughout the Americas, the Jerusalem Oak expels roundworms, hookworms and tapeworms and is especially useful for children. More scientific studies need to be done to confirm the historical usage of this herb.

Pumpkin seeds: Pumpkin seeds have also been used historically as a folk medicine for tapeworms and roundworms. In order to be really effective enormous amounts must be eaten:

7-14 oz for children and up to 25 oz for adults. Mash them and mix with juice. Two or three hours afterwards, take castor oil to clear your bowels.

Celiac Disease/Sprue/Gluten Intolerance

Celiac disease, also called celiac sprue or gluten sensitive enteropathy, is caused by an inability to properly digest foods containing gluten. Gluten is found in many grains including wheat, oats, rye, barley and spelt. The gliadin fraction in the gluten is believed to be the cause of the problem. In susceptible people gluten-containing grains damage the lining of the small intestine. Because people with celiac disease do not absorb nutrients well, they are likely to become malnourished. Celiac disease is chronic, has a genetic component and may affect several family members. It occurs about twice as often in women as in men and mainly affects people of northwestern European ancestry. It rarely occurs among people of African, Jewish, Mediterranean or Asian descent.

Some people have celiac disease without symptoms, while others can be very ill. Symptoms are characterized by recurring attacks of diarrhea, abdominal cramping, gas, weakness and steatorrhea (gray or tan fatty stools). Less common symptoms are arthritis and epilepsy, accompanied by folic acid deficiencies. Celiac disease is usually recognized early in childhood, but may disappear in adolescence and reappear later in adulthood. Anemia is common in people with celiac disease. A recent pediatric study showed that half of the people tested were anemic, and a small percentage of children with anemia really had undiagnosed celiac disease.[28]

Therapeutics consist of avoiding grains which contain gluten; that stops the irritation to the gut and allows it to heal. Many people find quick relief in three to four days, although complete healing of the intestines will take longer. Nutritional therapies may help you heal faster. If you don't feel better in four to six weeks, you need to investigate other possibilities that may be delaying your healing. Because of the malnourishment that celiac can cause, supplementation of nutrients is necessary.

It's not easy to completely change your diet to avoid all gluten-containing foods. Americans depend on a wheat-based diet and many other foods that have gluten-containing grains. In addition to obvious sources of gluten, many products have hidden sources. Salad dressings, some hot dogs, ice cream, bouillon cubes, chocolate and foods containing hydrolyzed vegetable protein may contain gluten. Fortunately, because of the many people who are intolerant of gluten, more gluten-free products are being introduced into the marketplace. Excellent breads, pastas, crackers, pancake mixes, cereals and cookies are now available. A new home test kit may soon be on the market which can help you determine which foods contain gluten. People with celiac disease are often lactose intolerant as well and should avoid dairy products, but this is usually temporary and resolves as they heal.

While celiac disease was previously thought to be rare, new studies indicate that many people without obvious symptoms have subclinical celiac disease.[29] According to analysis of IgG and IgA blood testing, an estimated 1 person in 258 may have celiac disease. Other studies place its incidence at 1 person in 1000 to 5000. What makes these figures so different? It appears that there are several types of gluten sensitivity: celiac disease, tropical sprue (caused by an infection or toxin and treated with antibiotics) and gluten sensitivity (caused by leaky gut syndrome). While symptoms may be similar for these problems, the treatments vary considerably. People with celiac disease must avoid gluten-containing foods for life. People with tropical sprue are generally treated with antibiotics and over time are able to use gluten-containing grains without further problems. People with wheat and/or gluten sensitivity often find that after avoidance of those foods for four to six months and a nutritional program that supports healing and friendly flora, they can then begin eating grain products without further problems.

FUNCTIONAL LABORATORY TESTING

1. Antigliadin antibody test and the rectal gluten challenge test: Diagnosis is usually made by excluding other possibilities and then performing an intestinal biopsy. New tests, such as the antigliadin antibody test (AGA) that determines celiac disease and the rectal gluten challenge test that determines gluten sensitivity, are useful without being invasive. Italian research has also found two additional tests to be accurate: the antijejunum antibody test (JAB) and the antiendomysium test (EMA). Both blood tests are not yet available in the United States.
2. Food allergy/sensitivity screening: Test for wheat, oats, rye, barley, gluten and gliadin; IgE and IgG antibody testing are all required. Gluten antibodies are positive in people with celiac disease and gluten intolerance. IgA levels are also higher in people with celiac disease, but a negative test does not always rule it out. Additional food sensitivities are likely.
3. Intestinal permeability screening: Test for leaky gut/ intestinal hyper-permeability.
4. Iron status/nutrient status: Celiac disease causes malabsorption of nutrients. Many people with celiac have iron deficiency anemia, low vitamin A, D, E and K, poor fat absorption and other mineral deficiencies. Several labs have tests which help determine which nutrients need supplementation.
5. Comprehensive digestive and stool analysis with parasitology: This test can be used to determine if there is an unidentified underlying cause of the celiac disease.
6. Rule out lactose intolerance: The lactose breath test determines lactose intolerance. Elimination of dairy products from your diet also helps determine if dairy products are contributing to your problem.

Healing Options

Diet: Avoid all gluten-containing grains and products that contain them, even in small amounts. Gluten-containing grains include wheat, rye, barley, oats, millet and spelt. It is essential to read all labels carefully and become an expert at reading between the lines.

Foods Which May Contain Hidden Gliadin

Texturized vegetable protein: It may contain *wheat* along with soy or corn.

Barley malt.

Starch (when listed in the ingredient list): If it doesn't say "cornstarch," it may contain gluten.

Desserts: Ice cream cones, ice cream which contains gluten stabilizers, most cakes, cookies and muffins.

Meats: Luncheon meats, hot-dogs and sausages may contain grains. Self-basting turkeys contain hydrolyzed vegetable protein which may contain gluten.

Cheese & dairy: Some processed cheeses contain wheat flour and/or oat gum.

Pasta: Most pasta is wheat. Semolina is a type of wheat which contains a high level of gluten. You can find gluten-free pasta at gourmet and health food stores.

Miscellaneous (some brands contain gliadin): Curry powder, most white pepper, dry seasoning mixes, gravy mixes and extracts, meat condiments, catsup, chewing gum, pie fillings, baked beans, baking powders, salad dressings, sandwich spreads, museli, cereals, instant coffee, breadcrumbs, vanilla and flavorings made with alcohol and most dips.

Digestive enzymes: Either pancreatic or vegetable enzymes can be used to enhance digestive function. Dosage: 1 to 2 with each meal.

Acidophilus/bifidobacteria: Probiotic flora enhance digestive function.

Gut-healing nutrients: Glutamine, gamma-oryzanol and N-Acetyl-D-Glucosamine are all healing to the intestinal lining. While no specific testing has been done on therapeutic use of these nutrients in people with celiac disease, clinical experience with celiac and other diarrheal illnesses indicates their usefulness.

Multivitamin with minerals: Zinc, selenium, folic acid, iron and vitamins A, B6, D, E and K have all been shown to be deficient in people with celiac disease. Get a good quality multiple vitamin with minerals. Look for a supplement that is hypoallergenic and contains no grains or dairy.

COLON

Common problems in the large intestine, also called the colon, include constipation, diarrhea, diverticular disease, irritable bowel syndrome, inflammatory bowel disease, ulcerative colitis, Crohn's disease, hemorrhoids, polyps and colon cancer. Proper functioning of the colon requires a high-fiber diet. The colon is home to trillions of beneficial bifidobacteria and other flora that ferment dietary fiber which, in turn, produces short-chained fatty acids, butyric acid, valerate, propionic acid and acetic acid. These short-chained fatty acids are the primary fuel of the colonic cells. Without adequate fiber, we starve the colonic cells and weaken the integrity of the colon. Butyric acid has been shown to stop the growth of colon cancer cells in vitro and is used clinically to heal inflamed bowel tissue.

The colon's main function is the reabsorption of nutrients and water into our bodies and the elimination of toxic wastes. Adequate intake of liquids is essential for good colon health. Water, juices, herbal teas and fresh fruits contain liquids that are hydrating to the stool. Coffee, alcohol and soft drinks are more diuretic, which increases our need for healthful fluids.

Constipation

Constipation affects 4 million Americans each year. Physicians write more than a million prescriptions for constipation annually, and we spend $725 million a year on laxatives. Constipation is a big problem! People are constipated when they strain to have a bowel movement, have hard stools, infrequent or incomplete bowel movements, discomfort or a perception that bowel habits are different from usual. Some people feel fatigue, achiness and mental sluggishness from constipation; others get headachy. Constipation affects women twice as often as men and is more common in people over age 65. Although aging is commonly listed as a cause of constipation, it is due more to the results of lifestyle. Elderly people often eat low-fiber foods, rely on packaged and prepared foods, take medications that interfere with normal bowel function and have decreased mobility. Each of these factors by itself increases the risk of chronic constipation and is not necessarily a progression of aging.

Many other factors can be the underlying cause of constipation. Hormones play a role. Women often notice that their bowel habits change at various times in their menstrual cycles. Pregnancy is a common, but temporary, cause of constipation, which can also be caused by an underactive thyroid. Some diseases can affect the body's ability to have bowel movements. Parkinson's disease, scleroderma, lupus, diabetes, kidney disease and certain neurological or muscular diseases like multiple sclerosis can cause constipation. Colon cancer can also cause it. Neurological problems such as injuries to the spinal column, tumors that sit on nerves, nerve disorders of the bowel and certain brain disorders are other causes.

Bowel movements should be painless. If you experience pain, see your physician. You may have a structural abnormality, fissure, hemorrhoid or more serious problem. Pain during bowel movements can cause a muscle spasm in the sphincter which can delay a stool. Magnesium helps relieve and prevent muscle spasms.

Bowel transit time is a newer parameter of constipation that is widely used in preventive health care. Bowel transit time is the amount of time it takes for food to go from the mouth, through the digestive system and out in stool. Reports from the National Institutes of Health suggest that a "normal" range of bowel movements is from three to twenty-one each week.[30] While this may show what is "average," it is not a good indicator of what is "normal." It is normal to have one to three soft bowel movements each day. Optimal bowel transit time is 12 to 18 hours, so if you are only having three bowel movements each week, you have a transit time of 56 hours, which is way too long. Slow bowel transit times raise the risk of colon diseases and contribute to other health problems due to the reabsorption of toxins back into the body. If you haven't done the transit time self-test, now would be a good time!

Dennis Burkitt, MD, studied bowel habits of Africans living in small towns and large cities. He found that people who ate indigenous local foods had an average of a pound of feces each day, with 12-hour transit times. He found that people living in cities on Western diets only excreted 5 1/2 oz of stool each day, with average transit times of 48 to 72 hours. People on native diets had extremely low incidences of diseases common to Western civilization, such as appendicitis, diabetes, diverticulitis, gallstones, coronary heart disease, hiatal hernia, varicose veins, hemorrhoids, colon cancer and obesity. When these people moved into cities and ate a Westernized diet, they too developed these diseases. Dr. Burkitt attributed much of this disease to poor dietary fiber intake in a modernized diet.

For most people a diet high in fiber and fluids solves the problem. Be sure to drink 6 to 8 glasses of water, juices or herbal teas and eat at least five servings of fruits and vegetables each day. Brussel sprouts, asparagus, cabbage, cauliflower, corn, peas, kale, parsnips and potatoes contain high amounts of fiber. Make whole grains the rule and processed grains the exception. The addition of high-fiber cereals at breakfast can make a big difference. Legumes, like kidney, navy, pinto and lima beans, have a large amount of dietary fiber.

Make these dietary changes slowly. A quick change to a high-fiber diet can cause gas pains. As your body gets used to this new way of eating, it will adapt. Remember that the requirements for most of us are to double our daily fiber intake. Cut back on low-fiber foods, including meats, dairy products, pastries, candy, soft drinks and white bread.

Exercise helps relieve constipation by massaging the intestines. Many of my clients have found that regular exercise keeps their bowels regular.

Overuse of laxatives is common and contributes to your problems. Chronic use of laxatives causes the bowels to become lazy, and the muscles become dependent on laxatives to constrict. People often find they need more and more laxatives to have the same effect. Some laxatives can cause damage to the nerve cells in the wall of the colon. If you have used laxatives, you need to retrain your body to have bowel movements on its own. Try sitting on the toilet each morning for 20 minutes and relax. Over time your body will remember how to relax and function normally.

Pay attention to your body's needs. When your body gives you the signals that it's time to defecate, stop what you are doing and go to the bathroom. When you ignore your body's urges, the rectum gets used to being stretched and fails to respond normally. Feces back up into the colon, causing discomfort. If you dislike having a bowel movement at work, school or in a public restroom, readjust your attitude and get used to the idea. Everybody's doing it!

Healing Options

Double your fiber intake: Fresh produce, organically grown if possible, gives life to our cells. Eat a minimum of five fruits and vegetables every day. More is better. They are rich in fiber, vitamins, minerals, trace nutrients, fluids and vitality. Eat whole grains like whole wheat berries, oatmeal, millet, amaranth, quinoa, rye berries and brown rice. Eat whole grain cereals at breakfast; they can significantly boost your fiber intake in just one meal. Beans and peas are also loaded with healthful fibers. They are a low-fat protein source, and their soluble fiber and sitosterols help normalize cholesterol levels.

Psyllium seed husks: Psyllium seeds add bulk and water to stool, which allows for easy passage. Though not a laxative, psyllium seeds do regulate bowel function, are beneficial for both diarrhea and constipation and do not cause harmful dependencies. Build up gradually to one teaspoon of psyllium with each meal to avoid gas and cramping from sudden introduction of fiber. As your dietary fiber increases, you will probably find that you no longer need psyllium seeds.

Wheat bran/corn bran: Wheat and corn bran can be used in the same way as psyllium seeds. They add bulk and moisture to stool, which allow them to pass more easily. Use 1 tsp with meals.

Increase fluids: Drink at least 6 to 8 glasses of water, fruit juice or herbal teas each day. Coffee, alcoholic beverages and soft drinks have a dehydrating effect on the body.

Improve bowel habits: Ignoring your body's natural urge to defecate can cause constipation. Take time each morning to have a bowel movement. If you go when nature calls, it takes just a minute or two.

Stop using laxatives and enemas.

Improve bowel flora: Poor bowel flora causes the digestive system to move sluggishly. Use of antibiotics, hormones and steroid drugs, high stress levels and poor diet can also cause an imbalance of intestinal flora. Bifidobacteria help regulate peristalsis. Dosage: Acidophillus and bifidobacteria two to three times daily. If you are able to digest yogurt, it has a normalizing effect on the bowels and can be helpful for either constipation or diarrhea.

Magnesium: Magnesium helps keep peristalsis—rhythmic muscle relaxation and contraction—working by proper relaxation of muscles. Americans have widespread magnesium deficiency that contributes to constipation. According to recent studies, 75 percent of magnesium is lost during food processing, and 40 percent of Americans fail to meet the RDA levels for daily magnesium intake.[31] When magnesium deficiency or a calcium/magnesium imbalance is present, poor bowel tone can occur. On the other hand, too much magnesium can cause diarrhea. Dosage: 400-500 mg daily.

Lactose intolerance: People with lactose intolerance sometimes become constipated from dairy products. Avoid milk and dairy-containing foods for two to three weeks and see if there is a change. Read the section on lactose intolerance in Chapter 7 to find hidden sources of lactose. Take the lactose breath test to determine if you are intolerant.

Medications: Many medications can cause constipation: pain relievers, antacids that contain aluminum, antispasmodic drugs, antidepressants, tranquilizers, iron supplements, anticonvulsants, diuretics, anesthetics, anticholinergics, blood pressure medication, bismuth salts and laxatives. If you noticed that constipation occurred suddenly after you began to take a new medication, discuss it with your doctor.

Begin an exercise or movement program.

Food sensitivities, dysbiosis, leaky gut syndrome: People with chronic constipation who do not respond to diet, fiber, liquids and exercise should have digestive testing to see if dysbiosis, food allergies or parasites are the underlying problem.

Vitamin C: Vitamin C can help soften stool. The amount varies depending on individual needs. Use a vitamin C flush to determine your daily needs.

Biofeedback: Biofeedback has been used successfully to treat constipation in people who have problems relaxing the pelvic floor muscles.[32]

Diarrhea

Diarrhea is a symptom, not a disease. If you have chronic diarrhea, it's important to find the underlying cause. Chronic diarrhea can be the result of drugs, diverticular disease, foods or beverages that disagree with your system, infections—bacterial, fungal, viral or parasitic—inflammatory bowel disease, irritable bowel syndrome, malabsorption of fats, lactose intolerance, laxative use and abuse, malabsorption, contaminated water supply or cancer. People with gallbladder problems often experience diarrhea after a fatty meal. With careful questioning and laboratory testing, your physician will be able to find the cause. Once you have a diagnosis, you can decide how to approach the problem.

Diarrhea occurs when you have a bowel transit time that is too fast. Feces don't sit in the colon long enough for water to be absorbed back into your body so the stool comes out runny. (It's truly amazing how much water is usually absorbed through the colon.) If you have chronic diarrhea, you aren't getting the maximum benefit from foods because you aren't absorbing all the nutrients. Loss of fluids and electrolyte minerals can make us disoriented and weak. In infants, small children and the elderly, dehydration can be dangerous and can happen suddenly. It's important to replace lost fluids to prevent dehydration. Drink 8 to 10 glasses of water, fruit and vegetable juices or broths each day. Infants can be given a fleet enema, which is easily purchased at drug stores. Follow the directions in the package.

Most diarrhea is self-limiting. If you have severe abdominal or rectal pain, fever of at least 102°F, blood in your stool, signs of dehydration—dry mouth, anxiety, restlessness, excessive thirst, little or no urination, severe weakness, dizziness or light-headedness—or your diarrhea lasts more than three days, call your doctor. Be more cautious with small children, people who are already ill and the elderly.

Diarrhea is the body's way of getting rid of something disagreeable—food, microbes or toxins. You usually aren't very hungry when you have acute diarrhea. Many foods "feed" the bugs and you instinctively stop eating. The diet recommended for people with diarrhea is called the BRAT diet, which stands for bananas, rice, apples and toast. These foods are bland and binding. Soda crackers, chicken and eggs can also be eaten. You can make a pretty tasty rice pudding with apples, rice, eggs and cinnamon, which is also binding.

Many other substances can cause diarrhea, including an excess of vitamin C and magnesium. For instance, antacids which contain magnesium salts can cause diarrhea. Sorbitol, mannitol and xylitol are sugars found in dietetic candies and sweets that cause diarrhea in most people. Even in small amounts, they can cause diarrhea in people sensitive to them. Some people have the same reaction to fructose or lactose.

Functional Laboratory Testing

Prolonged diarrhea is a symptom that warrants thorough investigation. These are a few of the tests that may give you information about what's causing your problem.

1. Complete digestive and stool analysis with parasitology.
2. Hydrogen breath test for small bowel overgrowth.
3. Lactose breath test.
4. Elisa food sensitivity screen.

Healing Options

Healing options depend on the cause of the diarrhea. See the appropriate sections for complete healing options.

Eliminate dairy products: Lactose intolerance is a common source of diarrhea. Avoid milk and dairy-containing foods for two or three weeks to see if the diarrhea stops.

Food allergies and sensitivities: Diarrhea is a common symptom of food sensitivities and allergies. Blood testing can help you determine which foods you are reacting to. Or you can go on an elimination diet for one week and slowly reintroduce foods. Although a painstaking process, you can find most food sensitivities and allergies this way. For more complete instructions, see food sensitivities and the elimination/provocation diet in Chapter 7.

Golden seal: This herb is highly effective for treatment of acute diarrhea caused by microbial infection. Be sure to use golden seal in recommended dosages as it may also cause diarrhea if used in excessive amounts.

Acidophilus and bifidobacteria: These beneficial bacteria help normalize bowel function. They ferment fiber which produces short-chained fatty acids to fuel the colonic tissue. You can also take probiotic supplements to help prevent traveler's diarrhea.

Saccharomyces boulardii: This friendly yeast has been used successfully to prevent and treat diarrhea caused by antibiotics. It boosts levels of secretory IgA, which is a protective part of the immune system. Saccharomyces is often found in supplements that contain acidophillus and bifidobacteria.

Yogurt: Yogurt is a folk remedy for diarrhea. It contains active bacteria, *L. thermophilus* and *L. bulgaricus*, which help prevent and stop diarrhea.

Handwashing: The simple act of washing hands with soap can help reduce the incidence of on-going diarrhea. In a study done with mothers of children with prolonged diarrhea the mothers were simply asked to wash their own hands with soap and water before preparing food and eating and to wash their children's hands before eating and as soon as possible after a bowel movement. There was an 89 percent reduction in diarrhea in the hand-washing mom's group in comparison with the control group.[33]

Diverticular Disease

Diverticula are pea-sized pouches that have blown out of the intestinal wall, primarily in the colon. The underlying cause of diverticula formation is constipation. Soft, bulky stools easily pass through the colon and respond to peristaltic waves; hard, dehydrated stools are harder to push along, and the bowel wall has to work harder. As a result, the muscles in the colon thicken to help this abnormal situation, which results in greatly increased pressures within the bowel. Over time, this prolonged pressure can blow out portions of the bowel wall, causing diverticular pouches.

Thirty to forty percent of all people over the age of 60 have diverticular disease, which occurs more commonly in women than men and with increasing frequency with age. Eighty-five percent of the people who have diverticula are symptom-free, and it's not a problem unless the pouches become inflamed or rupture. They usually heal on their own, but sometimes require surgery. When they become infected, causing diverticulitis, there is pain and often a fever. It is usually at this point that a physician will order tests to discover diverticulitis and diverticulosis. These infections are treated with antibiotics and a soft-fiber diet. Once the inflammation resolves, a high-fiber diet is recommended. Diverticular pouches don't go away, but a high-fiber diet will prevent most further attacks. Repeated episodes of diverticulitis may require surgery. A disease of Western civilization, diverticular disease occurs rarely in people who consume a high-fiber diet.[34]

Healing Options

High-fiber diet: A high-fiber diet is of first and foremost importance for preventing the development and reoccurrence of diverticular disease. If you are recovering from a flareup of diverticulitis, begin with a soft-fiber diet. Cook vegetables until fairly soft, eat cooked fruits, use easy to digest grains like oatmeal and make vegetable soups with tofu. Foods with seeds (like strawberries, poppy seeds, sesame seeds, pumpkin seeds) can catch in your diverticula and cause irritation. Until healed, avoid seed foods.

Once you are feeling well, establish a high-fiber diet as a normal part of your life. Focus on fruits, vegetables, whole grains and legumes. Meat, poultry, fish and dairy products contain zero fiber and need to be eaten in moderation. Studies have shown that people eating a high-fiber, low-fat diet lower their risks of diverticular disease significantly. (Men who eat a high-red meat, low-fiber diet have even higher incidences.[35]) It may take you some time to get accustomed to a high-fiber, low-fat diet, but it will be worth the effort. The benefits reach further than your digestive tract, lowering your risk factors for cancer, heart disease and diabetes. Be certain to drink plenty of water and other healthy beverages.

Acidophilus/bifidobacteria: Friendly flora can help fight the infection while it's active and protect you from future infection. Dosage: 1 capsule two to three times daily for prevention; 2 capsules three times daily during flareup.

Gamma-oryzanol: While studies of gamma-oryzanol, a compound in rice bran oil, were not directly involved with diverticulitis, gamma-oryzanol is known to have a healing effect on the colon (see discussion under Gastric Ulcers and Gastritis above). Rice bran oil is available in the United States at some gourmet stores. Dosage: 100 mg three times daily for a trial period of three to six weeks.

Glutamine: The digestive tract uses glutamine, the most abundant amino acid in the body, as a fuel source and for healing. Studies have not been done with diverticulitis, but glutamine is regenerative for the colon in general. Dosage: Begin with 8 g daily for four to eight weeks.

Evening primrose oil: Evening primrose oil increases the levels of prostaglandin E2 series, which promote healing and repair. Dosage: 1000-2000 mg three times a day.

Aloe vera: Aloe vera, which contains vitamins, minerals and amino acids, has been used by many cultures to heal the digestive tract. Its anti-inflammatory properties are soothing to mucous membranes, and it has been shown to reduce pain. It also stimulates the immune system, increasing white blood cell activity and formation of T-cells, and contains enzymes that help break down dead cells and toxins. It also reduces bleeding time, which is important with ruptured diverticula.

Slippery elm bark: Slippery elm bark has demulcent properties and is gentle, soothing and nourishing to mucous membranes. Drink as a tea, chew on the bark or take in capsules. Recipe for tea: Simmer 1 tsp of slippery elm bark in two cups of water for 20 minutes and strain. Sweeten if you want and drink freely; it can be used in large amounts without harm. Dosage: 2 to 4 capsules three times daily.

Multivitamin with minerals: Take a good supplement.

Irritable Bowel Syndrome

Irritable bowel syndrome (IBS) affects 10 to 20 percent of all American adults and is the most common gastrointestinal complaint, although most people who have it never seek a physician's help. Over the years IBS has had a variety of names: spastic colon, spastic bowel, mucous colitis, colitis and functional bowel disease. Associated symptoms are abdominal pain and spasms, bloating, gas and abnormal bowel movements. Diarrhea alternating with constipation is the most common pattern. Bowel movements usually relieve the discomfort. People with irritable bowel syndrome do not have any obvious changes in bowel structure or other serious complications and rarely require hospitalization. Nonetheless, IBS can significantly restrict one's lifestyle. Twice as many women as men seek medical help for IBS, and it often appears in the teen years or early adulthood.

Women may experience a flareup in their symptoms around their menstrual period. The most common symptom associated with menstruation is pain. Because chronic pelvic pain can lead to hysterectomy, physicians need to be clear that IBS is not causing the pain. When IBS is the culprit, hysterectomy won't eliminate it.

Anemia, weight loss, rectal bleeding, and fever are not symptoms of irritable bowel syndrome. Bowel changes accompanied by these symptoms need to be checked out by a physician to discover the cause.

The causes of irritable bowel syndrome are multifactoral. Many studies are flawed because they only test one factor at a time and generalize about results. For instance, some people with IBS improve when they eliminate wheat from their diets, while the majority won't notice any benefit. Unfortunately, there's no single cause for IBS, but hopefully we can find the causes for each person and work with them individually in response to their biochemical uniqueness.

Low-fat, high-fiber diets are best. Eat small meals, and chew food well. Dietary recommendations need to be tailored to your personal reactions. It is commonly advised to avoid alcohol and monitor sugar intake, coffee, beans and cabbage family foods (broccoli, brussel sprouts, cauliflower) because they can be difficult to digest. You need only avoid those foods if they bother you.

Food allergies and intolerances, medication, stress, hormone changes, low fiber diets and infection all play a role in IBS. Food sensitivities are found in half to two-thirds of people with IBS and are more prevalent in people who have allergies or come from allergic families.[36] The most common foods that trigger IBS are wheat, corn, dairy products, coffee, tea, citrus fruits and chocolate.

Undiagnosed lactose insufficiency is often the cause of IBS. In a recent study of 242 people it was found that 43 percent had total remission of IBS when they excluded dairy products from their diet, and another 41 percent had partial improvement.[37] Taking the lactose hydrogen breath test is a valuable way to discover who would benefit from a lactose-free diet. You can also discover this by avoiding all dairy foods and products which contain dairy foods for at least two weeks to see how you feel. If you have only moderate improvement, other foods may also be playing a role in your symptoms.

Lactose is not the only sugar to cause problems. Our cells use single sugar molecules (monosaccharides), but many foods contain two-molecule sugars (disaccharides) which must be split. New research suggests that many people are unable to split mannitol, sucrose, sorbitol, fructose and other disaccharides, and a high percentage of people with IBS are intolerant of one or more of these sugars. The result is diarrhea, gas and bloating. These people find that fruit, especially citrus fruit, aggravates their symptoms. Hydrogen breath tests are being used experimentally to diagnose this, but are not on the market at the present. You can do a self-test by avoiding all fruit and sugar for at least ten days. Be sure to read labels carefully and avoid any product which contains glucose, sucrose, malt, maltose, corn syrup, fructose, brown sugar, honey, maple syrup, molasses and lactose. You'll find that sugar is everywhere, but if disaccharides are the cause of your IBS, it is worth the time and trouble. If sugars and fruits make you feel worse, do the self-test and a blood or stool test for candida infection. People with candida often feel worse after eating sugars and fruits.

Recent technology has provided new insight into the causes of IBS. High levels of methane shown in hydrogen breath tests are associated with IBS and indicate an infection in the small intestine, most commonly *Clostridium difficile*. (See more on lab testing in Chapter 7.)

Parasites and candida overgrowth are overlooked causes of irritable bowel syndrome. Leo Galland, MD, has found that Giardia was responsible for problems in nearly half of his patients with IBS.[38] Even benign pinworms can cause severe colonic cramping at a certain stage of their lifecycle. Ask your doctor to order a comprehensive digestive and stool analysis with parasitology to determine if parasites or candida are making you sick.

Antibiotics are well-known causes of temporary diarrhea and GI problems. Steroid medications can also affect the balance of flora. The good flora are eliminated, especially in people who are on repeated doses of antibiotics, which allows other microbes to dominate the intestinal tract. Acidophillus, bifidobacteria and *Saccharomyces boulardii* supplements can help restore intestinal balance.

Irritable bowel syndrome has also been treated as a psychosomatic illness—"it's all in your head"—because there appears to be no other obvious reason for it. People with IBS often have reason to feel stressed-out, nervous and depressed about their condition. Although stressful situations often trigger IBS symptoms, stress itself doesn't cause IBS, and people with IBS have no greater incidence of depression than other people with chronic health problems. IBS can cause social embarrassment and is the second most common cause for missing work, after the common cold.

Functional Laboratory Testing

1. Comprehensive digestive and stool analysis with parasitology.
2. Lactose breath test.
3. Hydrogen breath test for methane levels.
4. Elisa food sensitivity testing—IgG.
5. Food allergy testing—IgE.
6. Intestinal permeability screening.

Healing Options

Fiber: Fiber and high-fiber diets are recommended for people with IBS. Focus on eating a minimum of five daily servings of fresh fruits and vegetables, plus whole grains, beans and peas. You can use high-fiber cereals to boost fiber content, but recent research indicates that wheat bran made the problem worse in 55 percent of cases, whereas it only improved symptoms in 10 percent of patients.[39] This is not surprising because a significant number of people with IBS have a hypersensitivity to wheat products.

If you want to add a fiber supplement, use psyllium seeds. Psyllium seeds were found to be more effective than wheat bran overall. They caused a decrease in abdominal bloating, whereas wheat bran increased bloating. In one study in which psyllium was given to people with IBS it improved several parameters by increasing the number of bowel movements per week, enlarging stool weight and speeding up transit times. No negative side effects were reported.[40]

Lactose intolerance: Lactose intolerance is often the underlying cause of IBS. Take the hydrogen breath test or eliminate all dairy products and products containing dairy from your diet for at least two weeks to help you determine whether lactose intolerance is contributing to your problem.

Gamma-oryzanol: Gamma-oryzanol, a compound found in rice bran oil, is a useful therapeutic tool for IBS. It acts on the autonomic nervous system to normalize production of gastric juice (see discussion under Gastric Ulcers and Gastritis above). Dosage: 100 mg three times daily for a trial period of 3 to 6 weeks.

Glutamine: Glutamine, the most abundant amino acid in the body, is used by the digestive tract as a fuel source and for healing IBS. Dosage: 8 g daily for a trial period of four weeks.

Peppermint oil: Peppermint oil is a muscle relaxant that is widely used in England for IBS. To get the oil into the intestines intact, use enteric coated peppermint oil. (The coating prevents it from dissolving in the stomach.) Dosage: 1 to 2 capsules daily between meals. During a spasm, you can rub a drop or two of the oil inside your anus with a finger.

Herbs: Chamomile, melissa (balm), rosemary and valerian all have antispasmodic properties. They help relieve and expel gas, strengthen and tone the stomach and soothe pain. Valerian, hops, skullcap and passion flower are all calming herbs, which can be found in a combination product. Antidepressant medication is often used by physicians for IBS; these gentle, effective calmatives may give you similar results. Use these herbs in capsules, tinctures and teas.

Ginger: Ginger, either fresh or powdered, helps relieve gas pains. It can be added to foods or used in tea. Within 20 to 30 minutes you'll be belching and/or passing gas which will relieve the discomfort. To make tea: take two or three slices of fresh ginger or 1/2 tsp dried ginger in one cup of boiled water. Combine it with other herbs, like peppermint or chamomile, to enhance the effect.

Multivitamin with minerals: Take a good supplement.

Behavioral therapies: Biofeedback, self-hypnosis and other relaxation techniques are widely used to help people with IBS. Stress often triggers bowel symptoms, and learning stress modification techniques can alter our reactions. If we don't react with alarm to a situation, our body doesn't sense it as stressful.

Inflammatory Bowel Disease, Crohn's Disease and Ulcerative Colitis

Inflammatory bowel diseases (IBD) share many of the symptoms of IBS, but they are very different problems. IBD involves inflammation of the digestive tract, which can occur anywhere from the mouth to the rectum. Symptoms include abdominal pain, cramping and diarrhea, which may be accompanied by fever, rectal bleeding, abdominal tenderness, abscesses, constipation, weight loss, awakening during the night with diarrhea and a failure to thrive in children. Symptoms come and go and can go into remission for months or years, and about half of the people with IBD have only mild symptoms. People with IBD often develop complications, which include inflammation of the eyes or skin, arthritis, liver disease, kidney stones and colon cancer. Of people with ulcerative colitis, 20-25 percent eventually require surgery due to massive bleeding, chronic illness, perforation of the colon or risk of colon cancer. Though the risk of developing colon cancer for people with widespread IBD is much higher than in the general population, it doesn't rise when only the rectum and distal end of the colon are involved.

The two most common types of IBD are ulcerative colitis and Crohn's disease. Most cases are diagnosed before age 40. IBD tends to run in families and is more prevalent among people of Jewish descent. IBD affects half a million Americans. Ulcerative colitis and Crohn's disease are similar but have different characteristics. Ulcerative colitis is a continuous inflammation of the mucosal lining of the colon and/or rectum. In the descending colon it is sometimes called "left-sided disease," and in the rectum, it is called distal disease, ulcerative proctitis or proctosigmoiditis. If sores are present, they are shallow, and it is generally milder and easier to treat in the rectum.

Crohn's disease can occur anywhere along the digestive tract, from mouth to rectum, but is most common in the colon and ileum near the ileocecal valve. It is sometimes called "right-sided disease." Frequent symptoms are fevers that last 24 to 48 hours, canker sores in the mouth, clubbed fingernails and a thickening of the GI lining which may cause constrictions and blockage. Inflammation develops in a skip pattern, a little here and a little there, and goes more deeply into the tissues than with ulcerative colitis. In later stages it can form abscesses and fistulas, little canals that lead to other organs or form tiny caves. If you require surgery for Crohn's disease, it is important to know which part of the intestines were removed and which nutrients may have inadequate uptake (see the absorption chart in Chapter 7).

IBD is considered an autoimmune disease (your body begins attacking itself). The causes are multifactoral and have produced much debate. Four theories are that IBD is caused by an infection, a hypersensitivity to antigens in the gut wall, an inflammation of the blood vessels which causes ischemia (a lack of blood supply to the tissues), and food sensitivities. These causes are interconnected and may all be present to some extent in everyone with Crohn's disease. Medical treatment for IBD consists of anti-inflammatory drugs, steroids and sometimes antibiotics. While these medications can often relieve symptoms of IBD, they carry their own risks. The good news is that effective natural therapies address each of the four theories, reduce the need for prescription medications and heal the bowel. Among the hundreds of patients with IBD that Drs. Jonathan Wright and Alan Gaby have seen, most have improved, many dramatically. The key to success appears to be getting people into remission. Once a flareup has died down, dietary therapies are highly successful in preventing a reoccurrence.[41]

Diet plays an important role in IBD. The incidence of IBD is growing rapidly in Western countries, but is rare in cultures where people eat a native diet. People who eat a high amount of sugars and low-fiber diets have a higher incidence of IBD, and there are correlations of IBD with cigarette smoking and eating fast foods. There is no one diet that helps all people with IBD, although the Elemental Diet and Haas Specific Carbohydrate Diet (details of the Haas program are discussed in the Healing Options) work well for many people. The Elemental Diet is comprised of synthetic foods you drink or are given through a tube. The Elemental Diet, which has resulted in a reduction of intestinal permeability as well as its symptoms, has been found to be as good as steroids in reducing inflammation in a flareup of Crohn's disease.[42] But there are problems with use of the Elemental Diet. It is unpalatable to many people and they won't drink it. Newer products which are tastier are coming on the market.

The Specific Carbohydrate Diet eliminates all simple sugars. As discussed under IBS, many people are unable to split disaccharide sugars (lactose, sucrose, maltose and isomaltose) into single molecule sugars. This may explain, in part, why the diet is so successful. The Specific Carbohydrate Diet also eliminates grains, which generally cause inflammation of the intestines in people with IBD. Leo Galland, MD, has found that the Specific Carbohydrate Diet works well for people with IBD.

Leaky gut syndrome (increased intestinal permeability) is prevalent in IBD. Often an infectious or parasitic condition underlies it. A flareup of symptoms commonly occurs with infections. The most common microbes involved are *E. coli*, staphylococcus, streptococcus, Proteus, *Mycoplasma pneumoniae*, *Clamydia psittaci*, *Clostridium difficile toxin*, and *Coxiella burnetii*. Bacterial infections occurred in one-quarter of all reoccurrence of IBD in these studies. Research from England implicates measles as a possible cause of Crohn's disease. They found measles virus in diseased parts of the colon. Swedish researchers find a high incidence of IBD in people who were exposed to measles in utero.[43] Mold sensitivity, allergies to candida and other types of fungus have also proven to provoke IBD symptoms. Although there is not much research on the yeast connection and IBD, clinicians have often found antifungal therapies to be useful. Friendly flora have been found to be dramatically out of balance in people with IBD, so use of probiotic supplements is highly recommended. Use of the comprehensive digestive and stool analysis with parasitology screening and intestinal permeability test will uncover many of these problems.

Food sensitivities play a significant role in IBD. About half of Dr. Wright and Dr. Gaby's patients report significant improvement with use of an elimination diet over a three-week period.[44] After this, they gradually add foods back into their diet to see which ones provoke bloating, pain, diarrhea, bleeding or other symptoms. One study found that 13 percent of children with IBD were allergic to cow's milk during infancy.[45] Lyndon Smith, M.D., believes that allergy to milk products doesn't disappear as we age, but rather the symptoms change in nature. We start by being colicky babies, toddlers with chronic ear infections, children with chronic strep throat and adults with headaches. IBD may be another manifestation of this progression. Blood testing for food allergies and sensitivities is informative.

Some people with Crohn's have flareups in a seasonal cycle which suggests an allergy component to the illness. While studies have shown that allergy is a factor in a small number of people, a survey of members of the National Foundation of Ileitis and Colitis showed that 70 percent of people with IBD listed other symptoms which were probably allergic. This led one researcher to say "inflammatory bowel disease is just another possible facet of allergy."[46]

IBD is not caused by emotional illness or psychiatric disorder, though the condition may cause emotional problems due to its chronic nature, painful episodes and lifestyle limitations. Prolonged treatment with steroid medications can cause side effects of depression, mania or euphoria.[47]

Because of bleeding and continued irritation, malabsorption of nutrients is often found in people with IBD. These same nutrients are often vital for repair, so the cycle worsens. Low serum levels of zinc, an important nutrient for wound repair, are often found in people with IBD. Folic acid helps repair tissue and prevents diarrhea. Prolonged bleeding can cause deficiencies of copper, iron, folic acid and vitamin B12.

People with IBD have an increased level of leukotrienes, produced by neutrophils, which increase pain and inflammation. Many natural substances can modulate these effects. For instance, Omega 3 fatty acids, found in cold-water fish, can reduce inflammation caused by leukotrienes and the arachidonic acid cascade.

There is a higher incidence of IBD in women who take oral contraceptives.[48] Women with a history of IBD or with family history of IBD may want to choose a different form of birth control.

Functional Laboratory Testing

1. Comprehensive digestive and stool analysis with parasitology.
2. Lactose breath test.
3. Elisa food and environmental sensitivity testing.
4. Intestinal permeability screening.
5. Nutritional analysis of blood and/or hair.
6. Antioxidant analysis.

Healing Options

Dietary: Eliminate simple sugars and fast foods (flareups occur more frequently when fast foods are eaten). Grains and dairy products often aggravate the condition.

Specific carbohydrate diet: Many people have found relief from using the Specific Carbohydrate Diet outlined in Elaine Gottschall's book *Food and the Gut Reaction* (reissued as *Breaking the Vicious Cycle)*.[49] Foods which are allowed are beef, lamb, pork, poultry, fish, eggs, natural cheeses, homemade yogurt, fruits, nuts, pure fruit juices, weak coffee, weak tea, and peppermint and spearmint teas. Also allowed are corn, soy, safflower, sunflower and olive oils. No grains, dairy products, legumes or potatoes, yams or parsnips are allowed on the diet. No sugars or alcoholic beverages are consumed. For delicious recipes and more detail, read the book.

This diet is beneficial because it eliminates most foods that cause sensitivities—grains and dairy products. Similar to the candida diet, it helps restore intestinal balance. While going on the diet alone may be effective, it is most effective after laboratory testing has determined your unique biochemistry. For example, if you find that you are not sensitive to grains or dairy products, you'll be able to include them in your food plan. Most foods can be added back into your diet within six months, while fixed sensitivities must be avoided long-term.

Lactose intolerance: Hydrogen breath testing or elimination of all dairy products and foods containing dairy from your diet for at least two weeks can help determine whether lactose intolerance is contributing to your problem.

Food sensitivities: Food sensitivities play a significant role in ulcerative colitis and Crohn's disease, occurring approximately half the time. The most common offenders are dairy products, grains and yeast, followed in frequency by egg, potato, rye, coffee, apples, mushrooms, oats and chocolate. Some people are sensitive to more than one food. Going on an elimination diet/cleansing diet is a simple way to eat foods which are less likely to trigger symptoms. The slow addition of new foods in the challenge/provocation stage will give you an idea of which foods make you feel worse. You can also do a blood test for food sensitivities.

Omega 3 fatty acids: Omega 3 fatty acids are found in cold water fish and have been used to reduce inflammation in rheumatoid arthritis, psoriasis and ulcerative colitis by reducing the production of leukotrienes. Salmon, mackerel, herring, tuna, sardines, and halibut are all excellent sources of EPA/DHA oils. Eating these fish several times a week can supply your body with these essential fats. Brown and red seaweeds also provide generous amounts of Omega 3 oils, but carrageenan, an extract from seaweed, may increase the inflammation in the colon. While carrageenan is used in animals to produce IBD, in humans the research is not yet clear. To be on the safe side, avoid red and brown seaweeds.

You can also take 1 to 3 capsules of EPA/DHA oils daily. In a recent study, it was found that use of Max/EPA decreased disease activity by 58 percent over a period of 8 months. No patient worsened, and 8 out of 11 were able to reduce or discontinue use of medication.[50] The dosage was 15 capsules of Max/EPA, which contained 2.7 g of EPA and 1.8 g of DHA, per day.

Flax seed oil also contains Omega 3 oils plus Omega 6

oils. Omega 6 fatty acids also help reduce inflammation. Flax oil comes in a liquid or in capsules and must be kept refrigerated to prevent rancidity. Evening primrose and borage oil are excellent sources of Omega 6 fatty acids.

Quercetin: Quercetin, the most effective anti-inflammatory bioflavonoid, can be used to reduce pain and inflammatory responses and control allergies. Dosage: 500 to 1000 mg three to four times daily.

Acidophilus/bifidobacteria/Saccharomyces boulardii: Imbalance of friendly flora allows for proliferation of pathogenic microbes, such as candida, Bacteroides, Citrobacter and more. It is believed that dysbiosis is a primary cause of IBD in many cases. Use of probiotic supplements can help restore balance of intestinal flora. Dosage: 1 to 2 capsules three times daily 1/2 hour before meals.

Gamma-oryzanol: Gamma-oryzanol, a compound found in rice bran oil, is a useful therapeutic tool for gastritis, ulcers and irritable bowel syndrome (see discussion under Gastric ulcers and Gastritis above). Dosage: 100 mg three times daily for a trial period of three to six weeks.

Glutamine: Glutamine is the most abundant amino acid in our bodies. The digestive tract uses glutamine as the primary nutrient for the intestinal cells, and it is effective for healing stomach ulcers, irritable bowel syndrome and ulcerative bowel diseases. Douglas Wilmore, M.D. has been using high doses of glutamine to heal digestive tracts in people with IBD who have had surgery in which part of the colon was removed. When only a short portion of the colon remains, people develop chronic diarrhea, a condition called short bowel syndrome. With a high-fiber, high-glutamine diet and short-term use of growth hormones, Dr. Wilmore is able to help normalize bowel function.[51] Dosage: Begin with 8 g daily for a trial period of four weeks.

Butyrate enemas: Butyrate is the preferred fuel of the colonic cells. It is produced when fiber in the colon is fermented by intestinal flora, predominantly bifidobacteria. A few studies have shown that butyrate enemas, taken twice daily, helped heal active distal ulcerative colitis.[52]

Demulcent herbs: Demulcent herbs—marshmallow, slippery elm, acacia, chickweed, comfrey, mullein and plantain—are beneficial and soothing to the intestinal membranes and help stimulate mucus production. All are gentle enough to be used at will; try them in tincture, capsule or tea form.

Multivitamin with minerals and antioxidant nutrients: Due to general malabsorption and poor dietary habits in people with Crohn's disease and ulcerative colitis, it is wise to add a good-quality multivitamin with minerals to your daily routine. Deficiencies of many nutrients have been found in people with IBD: calcium/magnesium, folic acid, iron, selenium, vitamins A, B1, B2, B6, C, D, E and zinc. Because some researchers feel that oxidative damage plays a significant role in IBD,[53] the supplement should contain adequate amounts of antioxidant nutrients: at least 10,000 IU beta-carotene or other carotenoids, 400 IU vitamin E, 250 mg vitamin C, 200 mcg selenium, 5 mg zinc, plus other nutrients. It may also contain CoQ10, glutathione, NAC, pycnogenol, superoxide dismutase (SOD) and other antioxidants. It is best to buy a supplement that is free of foods, herbs and common allergens.

Hemorrhoids

About half of Americans over the age of 50 have hemorrhoids. They are not life-threatening or dangerous, just unpleasant. They occur when blood vessels in and around the anus get swollen and stretch under pressure, similar to varicose veins in the legs. They are found either inside the anus (internal hemorrhoids) or under the skin around the anus (external hemorrhoids). Internal hemorrhoids may become so swollen that they push through the anus. When they become irritated, inflamed and painful, they are called protruding hemorrhoids.

Straining during bowel movements is a common cause of hemorrhoids. The most common symptom is bright red blood with a bowel movement. Hemorrhoids are also common but temporary during pregnancy. Hormonal changes cause the blood vessels to expand. During childbirth extreme pressure is put on the anus. Hemorrhoids also occur in people with chronic constipation or diarrhea. Sitting for long periods, heavy lifting and genetics are other influential factors. In most cases, hemorrhoids go away in a few days. If you have bleeding that lasts longer, have your doctor examine you to rule out a more serious problem.

A high-fiber diet with plenty of fluids—water, fruit juices and herbal teas—helps prevent hemorrhoids since fiber and fluids soften stool so they pass through easily. No straining with bowel movements means less pressure on the blood vessels near your anus. So increase your intake of fruits, whole grains, legumes and vegetables, especially those containing the most fiber: asparagus, brussel sprouts, cabbage, carrots, cauliflower, corn, peas, kale and parsnips. Eating a high-fiber breakfast cereal significantly increases your fiber intake.

Hemorrhoids generally don't itch. If your anus itches mainly at night, you might have pinworms. The best time to check for them is at night while you itch. Place a piece of tape around your finger, sticky side out. Put the tape on your anus, pull it off and check for worms, which look like moving white threads. If you are checking one of your children, you can use the tape method or just look. Another cause of rectal itching is called pruritus ani which can be caused by food sensitivities, contact with irritating substances (laundry detergent or toilet paper), fungi, bacterial infection, parasites, antibiotics, poor hygiene or tight clothing. If you have hemorrhoids, you might find relief from the following suggestions.

PREVENTION

Diet: A high-fiber diet usually prevents hemorrhoids and allows them to heal. Increase your intake of fruits, vegetables, whole grains and legumes and drink plenty of fluids, including water, fruit juices and herbal teas.

Psyllium seed husks: Psyllium seeds add bulk and water to stool, which allows for easy passage. They regulate bowel function and can be beneficial for both diarrhea and constipation. Because they are not a laxative, they do not cause harmful dependency. Gradually build up to 1 tsp of psyllium with each meal to avoid gas and cramping from the sudden introduction of fiber. As your dietary fiber increases, you will probably find you no longer need to take psyllium seeds.

Wheat bran/corn bran: Wheat and corn bran can be used in the same way as psyllium seeds to add bulk and moisture to stool, allowing them to pass more easily. Take 1 tsp with meals or eat a high-fiber breakfast cereal. Be sure to add fiber to your diet slowly to prevent any gas or discomfort.

Acidophilus/bifidobacteria: Poor bowel flora causes the digestive system to move sluggishly. Use of antibiotics, hormones or steroid drugs, high-stress levels and poor diet can cause an imbalance of intestinal flora. Take acidophilus and bifidobacteria two to three times daily to help regulate peristalsis. If you are able to digest yogurt, it also has a normalizing effect on the bowel and can be helpful for either constipation or diarrhea.

Magnesium deficiency: Americans have widespread magnesium deficiency that contributes to constipation. According to recent studies, we lose 75 percent of magnesium during food processing, and 40 percent of Americans fail to meet the RDA levels for daily magnesium intake. One of the many functions of magnesium is the proper relaxation of muscles, and peristalsis is comprised of rhythmic muscle relaxation and contraction. When magnesium deficiency or a calcium/magnesium imbalance is present, poor bowel tone can occur.

On the other hand, too much magnesium can cause diarrhea. Dosage: 400-500 mg daily.

Lactose intolerance: People with lactose intolerance can become constipated from dairy products (see discussion in Chapter 7).

Hormone changes: Women often notice that their bowel habits change at various times in their menstrual cycle. Pregnancy is a common, but temporary, cause of constipation and hemorrhoids. An underactive thyroid can also cause constipation.

Vitamin C: Vitamin C can be used to help soften stool. The amount needed depends on your individual needs. Use a vitamin C flush to determine your daily need for vitamin C (see Chapter 9).

Healing Options

Change your bathroom habits: In many countries people squat to relieve themselves. A squatting position on the toilet takes pressure off the rectum and can help during a flareup of hemorrhoids. (You may feel a little silly, but who's watching!) Also wipe gently with soft toilet paper. It may help to wash your anal area with warm water after each bowel movement, or if you have a bidet, now is the time to use it.

Salves: Salves can soothe inflamed tissues. Spread vitamin E oil, comfrey or calendula ointment or golden seal salve gently on the anus with your fingers. Witch hazel is also soothing to hemorrhoidal tissue. Put some on a cotton ball and press gently. Repeat treatments several times daily.

Sitz-baths: Sitz-baths are an old-fashioned remedy for hemorrhoids that are still in favor with the medical profession. Place 3 to 4 inches of warm water in the bathtub and sit in it for 10 minutes several times daily. You can improve the results by adding 1/4 cup Epsom salts or healing herbs—chamomile, chickweed, comfrey, mullein, plantain, witch hazel and yarrow are all healing and soothing to mucous membranes. Most of these are weeds and may even be growing in your yard. (Comfrey is a very easy herb to grow; just put it in a place where it can spread. It helps with wound healing of any sort and is also soothing for colds and lung problems.) Bring a large pot of water to a boil. Steep 1 to 2 cups of fresh herbs or 1/2 cup of dried herbs until cool, strain and add to bathwater.

Horse chestnut: Horse chestnuts, also called buckeye, help tone blood vessels, improve their elasticity and reduce inflammation. They can also be used in a sitz-bath. Chop up 2 cups of horsechestnuts, add to boiled water, strain and add infusion to bathwater. Sit in bath twice daily for 10 to 15 minutes. You can also take 500 mg of the bark orally three times daily. Horse chestnut salves are also available.

Butcher's broom: Butcher's broom helps strengthen blood vessels and improves circulation. Dosage: 100 mg extract three times daily.

Vitamin E: Vitamin E helps bring oxygen to the tissues and promotes healing. You can use it topically or take it internally. Dosage: 400-800 IU of d-alpha tocopherol and mixed tocopherols daily.

Vitamin C and bioflavonoids: Vitamin C and bioflavonoids increase capillary and blood vessel strength so they don't rupture easily. Bioflavonoids are also essential to collagen formation and elasticity of blood vessels. Berries of all types and cherries have high amounts of protective bioflavonoids. Dosage: 500-2000 mg daily plus 100-1000 mg bioflavonoids, which can usually be purchased in a single supplement.

Chapter 13

Natural Therapies for the Diverse Consequences of Faulty Digestion

"When one comes into a city to which he is a stranger, he ought to consider its situation, how it lies as to the winds and the rising of the sun; for its influence is not the same whether it lies to the north or to the south, to the rising or to the setting sun. . . . From these things he must proceed to investigate everything else. For if one knows all these things well, or at least the greater part of them, he cannot miss knowing, when he comes into a strange city, either the diseases peculiar to the place, or the particular nature of the common diseases. . . ."

HIPPOCRATES, *ON AIRS, WATERS, AND PLACES*, C. 400 B.C.

THIS CHAPTER DISCUSSES functional approaches to arthritis, eczema, psoriasis, migraines, fibromyalgia, scleroderma, chronic fatigue syndrome and food and environmental sensitivities. All these seemingly diverse medical problems are rooted in digestive conditions like leaky gut syndrome, parasites and dysbiosis. When the intestinal lining heals and intestinal flora regain balance, these conditions often improve dramatically.

For each health condition I have provided general information about the disease, recommendations for functional laboratory testing and healing options, with the most important ones discussed first. With careful investigation and patience you may find the underlying conditions that influence how you feel. You won't need to follow every remedy, but many of them can be found in combination products. You will note that although these health conditions are different, many of the healing options are the same.

Of course, you'll need to clean up your diet, correct any nutrient imbalances, get regular exercise and find time to nurture yourself. If at first you don't find major improvement, keep working at it. You may not have found the best remedy or combination of therapies on the first try. Patience and perseverance bring the best results. It takes time to resolve chronic illnesses.

Arthritis

Arthritis refers to over a hundred diseases that cause inflammation of the joints. The old-fashioned term for arthritis is rheumatism, and today physicians who specialize in arthritis are called "rheumatologists." Arthritis affects 37 million Americans and accounts for 46 million medical visits per year.[2] The two most common types of arthritis are osteoarthritis and rheumatoid arthritis. Other common types include psoriatic arthritis, ankylosing spondylitis, gout, Lyme disease and Sjögren's syndrome. Each of these diseases has its own characteristics, but they all share the symptoms of pain and inflammation in joints. There are many causes for arthritis:

genetic, infections, physical injury, nutritional deficiencies, allergies, metabolic and immune disorders, stress and environmental pollutants and toxins. Several types of arthritis have well-documented associations with faulty digestive function, and osteoarthritis responds well to dietary changes. Rheumatoid arthritis, ankylosing spondylitis, lupus, Sjögren's disease and Reiter's syndrome may be caused by a combination of genetics, dysbiosis, food or environmental sensitivities and leaky gut syndrome.

The current drugs of choice for arthritis pain are nonsteroidal anti-inflammatories (NSAIDs). However, NSAIDs block the production of prostaglandins which stimulate repair of the digestive lining. This causes increased leaky gut syndrome. Use of NSAIDs in children with rheumatoid arthritis showed that 75 percent had gastrointestinal problems caused by the drugs.[3] And the more NSAIDs people take, the leakier the gut wall becomes, the more pain and inflammation follows, which sets up a continuously escalating problem. To make matters worse, many NSAIDs also have a negative effect on the ability of cartilage to repair itself. They block our body's ability to regenerate cartilage tissue by lowering the amounts of healing prostaglandins, glucosaminoglycans and hyaluronan and by raising leukotriene levels.

Other drugs commonly used to ameliorate the symptoms of arthritis also have well-known side effects. Natural therapies for arthritis reduce the need for such medications and their accompanying side effects. These natural therapies can be astonishingly effective.

The dietary connection between rheumatoid arthritis and food sensitivities was first noted by Michael Zeller in 1949 in *Annals of Allergy*. He found a direct cause and effect by adding and eliminating foods from the diet. He joined forces with Drs. Herbert Rinkel and Theron Randolph to publish a book called *Food Allergy* in 1951. Theron Randolph, MD is the father of a field of medicine called clinical ecology, which studies how our environment affects health. He found that people with rheumatoid arthritis who were not reacting to foods had at least one sensitivity to an environmental chemical.

Randolph sent questionnaires to over 200 of his patients with osteoarthritis and rheumatoid arthritis to assess how well treatments were working. Their responses showed that when they avoided food and environmental allergens, there was a significant reduction in arthritic symptoms.[4] Randolph also felt that other types of arthritis, including Reiter's syndrome, ankylosing spondylitis and psoriatic arthritis, have an ecological basis.

Since then other studies have been done on the relationship between food sensitivities and arthritis. In a study of 43 people with arthritis of the hands, a water fast of three days brought improvement in tenderness, swelling, strength of grip, pain, joint circumference, function and SED rate (a simple blood test that determines a breakdown of tissue somewhere in the body). When some of these people were tested with single foods, symptoms reoccurred in 22 out of 27 people.[5] In other studies the foods most likely to provoke symptoms after an elimination diet were, in order of most to least: corn, wheat, bacon/pork, oranges, milk, oats, rye, eggs, beef, coffee, malt, cheese, grapefruit, tomato, peanuts, sugar, butter, lamb, lemon and soy. Cereals were the most common food, with wheat and corn causing problems in over 50 percent of the people. In another study it was found that 44 out of 93 people with rheumatoid arthritis had elevated levels of IgG to gliadin. Among these 44 people, 86 percent had positive RA factors. In yet another study, 15 out of 24 people had raised levels of IgA, rheumatoid factor and wheat protein IgG with a biopsy of the jejunum. Six of the wheat-positive people and one of the wheat-negative people had damage to the brush borders of their intestines. The researchers felt that the intestines play an important role in the progression of rheumatoid arthritis. Increased intestinal permeability allows more food particles to cross the intestinal mucosa, which triggers a greater sensitivity response.[6]

The concept of food sensitivity and increased intestinal

permeability is gaining acceptance as more physicians see the clinical changes in their patients when they use this approach. Testing for food and environmental sensitivities, parasites, toxic metals, candidiasis, intestinal permeability and comprehensive digestive and stool analysis often provides an understanding of an underlying cause of the disease.

Candidiasis frequently plays a role in arthritis and is a possible aggravator in rheumatoid arthritis. Yeast in the gastrointestinal system may be the result of antibiotics, oral contraceptives, steroid medications, increased use of alcohol or sugar or a stressed immune system. Treatment of candida infections in the digestive system has improved rheumatoid symptoms in many cases.[7] To find out if it complicates your symptoms, do the self-test in this book as well as a blood or stool test.

If candida or another infection is present, your physician can recommend a variety of therapeutics, including both natural and pharmaceutical remedies. If you have increased intestinal permeability, nutrients such as glutamine, quercetin, gamma-oryzanol and beneficial flora can help heal the leaky cells. An elimination diet or fasting can significantly reduce joint inflammation, pain, and stiffness and increase mobility. By careful addition of foods over the course of three months you can see which foods cause symptoms to reoccur. Blood testing can significantly aid in this process because you have a much clearer idea of which foods you are sensitive to. No blood test is 100 percent accurate, so you still need to go through the dietary regimen. After a period of four to six months you will be able to tolerate most of the troublesome foods. Repeat testing at that time is advised.

After that, many people will be pain-free, while others may still have some arthritic symptoms. There is documentation in the literature about arthritis and deficiencies of nearly every known nutrient. Nutritional supplements often work well because people with arthritis due to leaky gut are generally malnourished. When the needed nutrients are supplied, the body can begin to balance itself. Though there are many additional nutritional and herbal products that help people with arthritis, no one thing works for everyone, so persist until you find the therapies that work best for you. Give each one at least a three-month trial before giving up on it. I remember Abraham Hoffer, MD telling about a patient at a conference many years ago. He had recommended the man take 1000 mg of vitamin C daily for his arthritis. The man took the vitamin C faithfully each day without any improvement. After a whole year, he suddenly became pain-free!

Specific recommendations for various types of arthritis—osteoarthritis, rheumatoid arthritis, psoriatic arthritis and ankylosing spondylitis—follow.

Osteoarthritis

Osteoarthritis is the most common type of arthritis. It's the kind we associate with aging, although nutritionally oriented physicians believe it has more to do with poor dietary habits and biochemical imbalances. Pain is usually the first symptom. The main characteristics are stiffness, achiness and painful joints that creak and crack. Stiffness may be worse in the morning and after exercise. Osteoarthritis begins gradually and usually affects one or a few joints, most commonly in the knee, hip, fingers, ankles and feet. As joints enlarge, cartilage degenerates. Eventually hardening leads to bone spurs. You lose flexibility, strength and the ability to grasp, accompanied with pain. Risk of osteoarthritis, especially arthritis in the knee, increases if you are overweight; losing weight helps.

FUNCTIONAL LABORATORY TESTING

1. Elisa allergy testing for foods.
2. Comprehensive digestive and stool analysis.
3. Intestinal permeability screening.
4. Candida testing, either separately or in CDSA.

Healing Options

Some of these suggestions will significantly help your arthritis; others may not help at all. Be patient and give whatever you try time to work. Try one or two at a time until you find a program that suits your body's unique needs and lifestyle.

Elimination/provocation diet: Follow the directions outlined in Chapter 7. For best results work with a nutritionist or physician who is familiar with food sensitivity protocols.

Nightshade diet: In the 1970s Norman Childers, a horticulturist, popularized the Nightshade Diet. Elimination of nightshade family foods helps only a small percentage of people with arthritis, but people who respond are usually helped a great deal. The nightshade foods are potatoes, tomatoes, eggplant and peppers (red, green, yellow, chile). An elimination diet of one week followed by a reintroduction of these foods provides a good test. Blood testing also picks up these sensitivities.

Multivitamin with minerals: People with arthritis are often deficient in many nutrients. Aging, poor diet, medications, malabsorption and illness all contribute to poor nutritional status. At least 17 nutrients are essential for formation of bone and cartilage, so it's important to find a supplement that supports these needs. Look for a supplement which contains 800-1000 mg calcium, 400-500 mg magnesium, 15-45 mg zinc, 1-2 mg copper, 10,000 IU vitamin A, 200 mcg selenium, 50 mg vitamin B6, and 5-10 mg manganese in addition to other nutrients. Follow dosage on bottle to get nutrients in the appropriate amounts.

Vitamin E: Twenty-nine people with osteoarthritis were given 600 IU of vitamin E or a placebo daily. Out of 15 who received vitamin E, 52 percent reported improvement. Another study showed no improvement in those with osteoarthritis who were given vitamin E supplementation of 1200 IU daily.[8] Dosage: Try 800 IU for two to three months. It is very safe and may help some people. Just be sure to buy d-alpha tocopherol, which is the natural and active form of vitamin E.

Vitamin C: Vitamin C is an essential nutrient for every anti-arthritis program. Vitamin C is vital for formation of cartilage and collagen, a fibrous protein that forms strong connective tissue necessary for bone strength. Vitamin C also plays a role in immune response, helping protect us from disease-producing microbes. Many types of arthritis are caused by microbes, which vitamin C helps combat. It also inhibits formation of inflammatory prostaglandins, helping to reduce pain, inflammation and swelling. Vitamin C is also an antioxidant and free-radical scavenger; free-radical formation has been noted in arthritic conditions. Dosage: 1-3 g daily in an ascorbate or ester form.

Glucosamine sulfate: Glycosamine sulfate and chondriatin sulfate are nutrients used therapeutically to help repair cartilage, reduce inflammation and increase mobility. Green-lipped mussels are a rich source of glycosaminoglycans. Use of glycosamine sulfate has no associated side effects.[9] Dosage: 500 mg three times daily.

Alfalfa: Alfalfa is a tried and true folk remedy for arthritis. Many people attest to its benefits, but more research is needed on it. Alfalfa is an abundantly nutritious food, high in minerals, vitamins, antioxidants and protein. Alfalfa may help because of its saponin content or its high nutrient and trace mineral content. It is widely used as a nutritional supplement in animal feed.

Dosage: 14-24 tablets in two or three dosages daily or grind up alfalfa seeds and take three tablespoons of ground seeds each day. You can mix them with applesauce, cottage cheese, oatmeal or sprinkle on salads. Another method is to cook 1 oz of alfalfa seeds in 3 cups of water. Do not boil them, but cook gently in a glass or enamel pan for 30 minutes, and strain. Toss away the seeds and keep the tea. Dilute the tea with an equal amount of water. Add honey if you like. Use it all within 24 hours. Yet another method is to soak 1 oz of alfalfa seeds in 3 cups of water for 12 to 24 hours. Strain and drink the liquid throughout the day.

Yucca: Yucca has been used by Native Americans of the Southwest to alleviate symptoms of arthritis and improve digestion. It's a rich source of saponins with anti-inflammatory effects. Studies have been done with both rheumatoid and osteoarthritis with significant improvement in 56 to 66 percent of the people who tried it. People taking yucca over one and a half years also had the additional advantage of improved triglyceride and cholesterol levels and reduction in high blood pressure, with no negative side effects.[10] Dosage: 2-8 tablets daily.

Omega 3 fatty acids/fish oils: Fish oils come from cold water fish and contain eicosapentaenoic acid (EPA) and dicosahexaenoic acid (DHA). The fish with the highest levels are salmon, mackerel, halibut, sardines, tuna and herring. These omega-3 fatty acids are essential because we cannot synthesize them and must obtain them from our foods. Fish oils inhibit production of inflammatory prostaglandin E2 series, inhibit cylcooxygenase and thromboxane A2, all of which come from arachidonic acid. Fish oils shift the production to thromboxane A3, which causes less constriction of blood vessels and platelet stickiness than thromboxane A2.

Research has shown fish oils are really helpful for some people with arthritis. Fish oil capsules reduce morning stiffness and joint tenderness. They produce moderate, but definite improvement in arthritic diseases at dosages from 8-20 capsules daily.[11] Similar results can be obtained by eating fish with high EPA/DHA two to four times a week. Because fish oils increase blood clotting time, they should not be used by people who have hemophilia or who take anticoagulant medicines or aspirin regularly. Dosage: It's easiest for most people to eat fish two to four times each week. High dosages in capsule form should be monitored by a physician.

Ginger: Ginger is an old ayurvedic remedy that was given to people with rheumatoid and osteoarthritis. It reduced pain and swelling in various amounts in 75 percent of the people tested, with no reported side effects over three months to two and a half years. Ginger can be used as an ingredient in food and tea or taken as a supplement. Dosage: 2 oz fresh ginger daily or 3000-7000 mg powdered ginger.[12]

Superoxide dismutase: Superoxide-dismutase (SOD) plays an important role in reducing inflammation and has been used alone, with copper, manganese or copper and zinc for various arthritic conditions. Oral SOD doesn't seem to work as well, except when used in a copper/zinc preparation. Some physicians are using SOD in injections. Wheat grass extracts of SOD can be purchased at health food stores. Most people who try them experience benefits, but there is little scientific research to date. Some veterinarians are using wheat grass SOD with arthritic animals with excellent results.

Bromelain: Bromelain is an enzyme derived from pineapple that acts as an anti-inflammatory in much the same way that evening primrose, fish and borage oils do. It interferes with production of arachidonic acid, shifting to prostaglandin production of the less inflammatory type. It also prevents platelet aggregation and interferes with the growth of malignant cells. Bromelain can be taken with meals as a digestive aid, but as an anti-inflammatory it must be taken between meals. Dosage: 500-1000 mg two to three times daily between meals.

Quercetin: Quercetin is a bioflavonoid that comes from oak trees. It is the most effective bioflovonoid in its anti-inflammatory effects; others include bromelain, curcumin and rutin. Bioflavonoids help maintain collagen tissue by decreasing membrane permeability and cross-linking collagen fibers, making them stronger. Quercetin can be used to reduce pain and inflammatory responses and for control of allergies. Dosage: 500-2000 mg daily.

Folic acid plus vitamin B12: In a recent study, people with osteoarthritis in their hands were given 20 mcg vitamin B12 plus 6400 mcg folic acid daily, and they reported a significant reduction in symptoms.[13] This is a tiny amount of vitamin B12 and a large amount of folic acid, which is nontoxic even at these high levels.

Cider vinegar: Jarvis in his book, *Vermont Folk Medicine*, popularized the use of apple cider vinegar and honey for the relief of arthritis. Vinegar and honey are alkalizing to the body, which helps dissolve calcification in the joints. Many people have found relief on this program, although there has been no scientific documentation.

Pulse electromagnetic fields: A small study was done on 27 people with osteoarthritis using low-pulsed electromagnetic fields at an extremely low frequency during 18 half-hour sessions. Halfway through the study 34 percent improved, at the end of treatment 36 percent improved, and one month later 47 percent had improved. This was significantly different than the control group—8, 10 and 14 percent.[14] This treatment has no known negative effects and is certainly worth trying.

Rheumatoid Arthritis

Rheumatoid arthritis is characterized by inflammation of joints most often in hands, feet, wrists, elbows and ankles with symmetrical involvement. It can start in virtually any joint. The onset may be sudden, with pain in multiple joints, or may come on gradually, with more and more joints becoming involved. Joints become swollen and feel tender and can degenerate and become misshapen. They are often stiffest in the mornings and also feel worse after movement. The rheumatoid factor (RF) is a blood test that will become elevated in most cases of rheumatoid arthritis. While it may get better or worse, once established it is nearly always present to some extent. Standard treatment is aspirin or NSAIDs and steroid medications.

Rheumatoid arthritis has a genetic component, often running in families. The gene marker HLA-DR4 is present in 50 to 75 percent of people with rheumatoid arthritis. Proteus, a bacteria commonly found in the digestive tract, typically doesn't cause illness, but when present in a person who is HLA-DR4 positive, it may trigger an autoimmune response which leads to rheumatoid arthritis. Some researchers believe that rheumatoid arthritis is the result of repeated episodes of Proteus infection.[15] Proteus infections can be treated with either natural or pharmaceutical therapy.

Food and environmental sensitivities, malabsorption, parasites, candida and leaky gut syndrome also play a role in rheumatoid arthritis. Some people with it are gliadin sensitive (sensitive to grain; see Celiac disease). It's hard to generalize or predict which of these factors will be found in each person, but usually one or more is present. Each of them needs to be investigated. Leaky gut is probably not a primary cause of rheumatoid arthritis, but long-term use of NSAID medication often makes it a factor.[16]

FUNCTIONAL LABORATORY TESTING

1. Elisa allergy testing for foods.
2. Comprehensive digestive and stool analysis.
3. Intestinal permeability screening. Stop use of NSAIDs for three weeks prior to test. You can also use the IP test to monitor your progress.
4. Heidelberg capsule testing for HCl status.
5. Methane breath test for small bowel bacterial overgrowth.
6. Liver function testing. People with rheumatoid arthritis are also shown to have reduced function in the detoxification pathways.

HEALING OPTIONS

Try one or two of these suggestions at a time, and you'll gradually find a program that suits your unique needs and lifestyle.

Food & environmental sensitivities: Explore the relationship between your arthritis and food and environmental sensitivities through laboratory testing and the Elimination/Provocation diet.

Elimination/provocation diet: Follow directions in Chapter 7. For best results work with a nutritionist or physician who is familiar with food sensitivity protocols.

Hypochlorhydria/low HCl levels: Low levels of hydrochloric acid (HCl) were found in 32 percent of people tested with rheumatoid arthritis. Half of these people had small bowel bacterial overgrowth. Thirty-five percent of patients with normal levels of HCl had small bowel bacterial overgrowth compared with none of the normal controls. Small bowel overgrowth was found most in people with active arthritic symptoms.[17]

Dosage: Varies from person to person. To find the optimal amount for you, begin by taking one HC1 tablet with each meal for one day. The next day, take two tablets with each meal, gradually increasing the dosage. At some point you will feel a warm, burning sensation in your stomach; your optimal dose is one tablet less than this. If the burning sensation is uncomfortable, quickly drink a glass of milk or water with a teaspoon of baking soda.

Nightshade diet: Elimination of nightshade family foods helps only a small percentage of people with arthritis, but people who respond are usually helped a great deal. A one-week elimination diet with a gradual reintroduction of nightshade family foods offers a good test. Blood testing also picks up these sensitivities.

Multivitamin with minerals: People with arthritis are often deficient in many nutrients. So it's important to find a supplement that supplies at least 17 nutrients essential for formation of bone and cartilage. Look for a supplement which contains 800-1000 mg calcium, 400-500 mg magnesium, 15-45 mg zinc, 1-2 mg copper, 10,000 IU vitamin A, 200 mcg selenium, 50 mg vitamin B6, and 5-10 mg manganese in addition to other nutrients. Follow dosage on bottle to get nutrients in the appropriate amounts.

Vitamin E: In a study with vitamin E supplementation of 1200 IU daily, there was a decrease in pain in people with rheumatoid arthritis.[18] (For dosage, see discussion under Osteoarthritis.)

Vitamin C: Vitamin C is an essential nutrient for every anti-arthritis program. (For more on vitamin C, see discussion under Osteoarthritis.) Dosage: Take 1-3 g daily in an ascorbate or ester form.

Alfalfa: Alfalfa is a tried and true folk remedy for arthritis, but more research is needed on it. Dosage: Take 14-24 tablets in two or three dosages daily or prepare alfalfa seeds as discussed under Osteoarthritis.

Yucca: Yucca has been used by Native Americans of the Southwest to alleviate symptoms of arthritis and improve digestion. (For more on yucca, see discussion under Osteoarthritis.) Dosage: 2-8 tablets daily.

Gamma linolenic acid: Patients with rheumatoid arthritis were given 1.4 g daily of gamma linolenic acid (GLA) from borage oil. It significantly reduced their symptoms: swollen joints by 36 percent, tenderness by 45 percent, swollen joint count by 28 percent and swollen joint score by 41 percent (some people responded in more than one area).[19] Use of oil of evening primrose in the study group and olive oil for the control group showed that both oils helped reduce pain and morning stiffness. Several people were able to reduce use of NSAIDs, but none were able to stop the medication.[20] The modest results in this study were probably due to the use of NSAIDs with the oil of evening primrose. The same results could be obtained by use of evening primrose or borage oil alone. Dosage: 1400 mg.

Omega 3 fatty acids/fish oils: Research has shown that some people with rheumatoid arthritis have found it very helpful to eat fish high in Omega 3 fatty acids—salmon, mackerel, halibut, sardines, tuna and herring—or take fish oil supplements. (For more on Omega 3 fatty acids/fish oils, see discussion under Osteoarthritis.) Dosage: It's easiest for most people to eat fish two to four times each week. High dosages in capsule form should be monitored by a physician.

Glucosamine sulfate: Glycosamine sulfate is used therapeutically to help repair cartilage, reduce swelling and inflammation and restore joint function. (For more on glycosamine sulfate, see discussion under Osteoarthritis.) Dosage: 500 mg two to four times daily.

Ginger: Ginger is an old ayurvedic remedy that was given to people with rheumatoid and osteoarthritis. (For more on ginger, see discussion under Osteoarthritis.) Dosage: 2 oz fresh ginger daily in food or tea or 3000-7000 mg powdered ginger.

Superoxide dismutase: Superoxide dismutase plays an important role in reducing inflammation and has been used alone, with copper, manganese or copper and zinc for various arthritic conditions. (For more on superoxide dismutase, see discussion under Osteoarthritis.)

Bromelain: Bromelain acts as an anti-inflammatory in much the same way that evening primrose, fish and borage oils do. (For more on bromelain, see discussion under Osteoarthritis.) Dosage: 500-1000 mg two to three times daily between meals.

Quercetin: Quercetin is the most effective bioflavonoid in its anti-inflammatory effects. (For more on quercetin, see discussion under Osteoarthritis.) Dosage: 500-1000 mg two to four times daily.

DL phenylalanine: DL phenylalanine is an amino acid that is used therapeutically for pain and depression. It is effective for treating rheumatoid arthritis, osteoarthritis, low back pain and migraines. "D" is the naturally found form, and "L" is its synthetic mirror. The combination of DL slows down the release of the phenylalanine. It appears to inhibit the breakdown of endorphins, our body's natural pain relievers. Dosage: 400-500 mg three times daily.

Copper: Copper is involved in collagen formation, tissue repair and anti-inflammatory processes. People with rheumatoid arthritis often have marginal copper levels. Traditionally, copper bracelets have been worn to help reduce arthritic symptoms. W. Ray Walker, Ph.D. tested people who had benefited from copper bracelets by having them wear copper-colored aluminum bracelets for two months: 14 out of 40 participants deteriorated so much they couldn't finish the two months. More than half reported that their arthritis had worsened. Dr. Walker found that 13 mg of copper per month was dissolved by sweat, and presumably much of that was absorbed through the skin. Supplementation with copper increases levels of SOD.[21] Dosage: Wear copper bracelet or supplement with 1-2 mg daily in a multivitamin preparation. If you are working with a physician, you may temporarily add a supplement of copper salicylate or copper sebacate until copper levels return to normal levels.

Niacinamide: Most of the B-complex vitamins have been shown to reduce inflammation and swelling associated with arthritis. Dr. Kaufman recommends using niacinamide at a rather high dosage with excellent results. If you are going to try this, do so with your physician's supervision. High levels of niacinamide can be liver toxic. Dosage: 500-2000 mg daily.

Cider vinegar: Although there is no scientific documentation about apple cider vinegar and honey, many people report relief. Because vinegar and honey are alkaline, they may help dissolve calcification in the joints.

Breast implants: Silicone breast implants may cause rheumatoid-like symptoms in some women, although research is divided.[22] If you have rheumatoid arthritis and silicone or saline breast implants, it would be smart to be tested for silicone antibodies/allergies on an annual basis. Many women feel remarkably better once breast implants have been removed.

Psoriatic Arthritis

Psoriatic arthritis affects 5 to 7 percent of people with psoriasis who have joint pain as well as the more common skin and nail symptoms. There is not yet agreement on whether this is an advanced stage of psoriasis or a completely separate but similar disease. The arthritic condition closely resembles that of rheumatoid arthritis, although people with psoriatic arthritis usually have a negative rheumatoid factor. Fingers, toes and the low back are the most likely joints to be affected. Skin and joint symptoms may flare up or improve simultaneously. This disease can be mild, but it can also be severely deforming and disabling.

No one really knows what causes psoriatic arthritis, although it appears to have genetic, environmental and immunologic origins. The gene marker HLA-B27 is present in most people with this disease. Some researchers have concluded that chronic exposure to bacteria in the gut, tonsils and skin results in inflammation throughout the body, particularly in the skin and joints. The interaction of these bacteria with a specific gene type may cause an autoimmune reaction, which causes the early symptoms.[23]

Inflammation of psoriatic arthritis is involved with arachidonic acid pathways. A healthful diet plus essential fatty acids help reduce and prevent further inflammation. Evening primrose, borage and fish oils, bromelain and quercetin all work on these pathways.

Healing Options

See Rheumatoid Arthritis and Psoriasis.

Ankylosing Spondylitis

Ankylosing spondylitis is characterized by a progressive fusion of joints in and around the spine. It affects Caucasian men 90 percent of the time and typically becomes evident between the ages of 10 to 30. It starts off as low back ache, which is often worst in the mornings. Symptoms get progressively worse and spread from the lower back to the mid-back and up to the neck. The spine gradually becomes fused. Later, shoulders, hips and knees may be affected. Symptoms flare and subside. Secretory IgA levels are usually elevated in people with ankylosing spondylitis.

The role of dysbiosis in ankylosing spondylitis is the most researched and best understood of all the arthritic diseases. An infection in the intestines, usually of a bacteria called Klebsiella, causes a local immune reaction. The antibodies for Klebsiella fit the gene HLA-B27. While Klebsiella is normally not a disease-producing microbe, it appears in people with the HLA-B27 gene marker as if the antibodies produced to kill it attach to the HLA-B27 cells and destroy them, causing joint pain and inflammation. This concept of autoimmune disease may explain why some people get certain illnesses and others don't. It's the presence not only of a specific gene, but also of a microbe that activates destruction of the cells.

What begins as a local infection triggers an autoimmune disease. As the latest research shows, "An association between inflammatory bowel disease and enteroarthritis and the spondyloarthropathies has been known for a while . . . and it now seems evident that chronic gut inflammation is either associated with or is even the cause of chronicity of peripheral arthritis and the development of ankylosing spondylitis."[24]

While the gene marker HLA-B27 is present in 96 percent of people with ankylosing spondylitis, not all people with this gene marker have the illness. This marker is also present in 10 to 15 percent of the general population. Research shows 70 to 80 percent of people with ankylosing spondylitis have the Klebsiella bacteria in their stools. Yersinia, shigella and salmonella bacteria are also associated with this process and may contribute to the disease in people who do not have Klebsiella.[25]

It is important to make an early diagnosis of ankylosing spondylitis so that progression of the disease can be slowed or halted. Because it usually appears as a low back ache, many people seek chiropractic help or massage therapy or take anti-inflammatory medications. But such remedies can't correct dysbiosis in the intestinal tract. Because men commonly have low back pain, they often have irreversible damage before a correct diagnosis is made.

Leaky gut syndrome is present in people with ankylosing spondylitis. Unfortunately, NSAIDs are commonly used to treat ankylosing spondylitis, causing even greater intestinal permeability. This, in turn, causes more sensitivity to foods and environmental substances.

Functional Laboratory Testing

1. Secretory IgA.
2. Comprehensive digestive and stool analysis.
3. Elisa food and environmental sensitivity screening.
4. Intestinal permeability screening.

Healing Options

Treat the infection: Your physician can prescribe an appropriate antibiotic to treat the underlying infection. If you are working with a physician who specializes in natural therapies, he or she may suggest the use of colloidal silver, golden seal and/or grapefruit seed extract.

Bifidobacteria and acidophilus: Because treatment for the infection will alter your intestinal flora, take probiotic supplements of supportive healthy bacteria to reestablish them. Dosage: 1-2 capsules two to three times daily or 1/2-1/4 tsp powder mixed with a cool or room-temperature beverage taken between meals.

Follow the recommendations for osteo- and rheumatoid arthritis: Once the underlying infection has been treated and flora have been reestablished, try other healing options for your remaining symptoms.

CHRONIC FATIGUE SYNDROME

Fatigue is one of the most common complaints that bring people into a physician's office. Fatigue can be caused by nearly every illness and is part of the natural healing process. Excessive fatigue that lasts and lasts may be a sign of illness or of chronic fatigue syndrome. Also called CFIDS, CFS, myalgic encephalomyelitis, chronic Epstein-Barr virus and yuppie flu, chronic fatigue syndrome is a long-lasting, debilitating fatigue that is not associated with any particular illness. Although people have been fatigued for millennia, the term "chronic fatigue syndrome" was only coined in 1988. CFIDS affects about 3 people per 100,000, and symptoms last approximately two and a half years on average. Most people eventually return to normal health.[26]

People with chronic fatigue syndrome have by definition been extremely tired for at least six months for no obvious reason. CFIDS often begins with an infectious disease, like a flu, accompanied by fevers that come and go. There is often accompanying joint stiffness and pain, sore throat, cough, sleep disturbances, light sensitivity, night sweats and extreme exhaustion after the slightest exertion. It's common that a short walk or bit of exercise wipes out your energy for days afterwards. People with CFIDS share many common symptoms, but not everyone has all the same ones. Some people have the Epstein-Barr virus, or Cytomegalovirus, but others don't. Sometimes healthy people have high blood titers for these viruses and have no symptoms of CFIDS. It's possible that these viruses trigger CFIDS, but it's also possible that the low immune function in people with CFIDS increases their chances of catching a wide variety of infectious illnesses.

Many people with CFIDS cannot hold down a job and many are depressed because the fatigue is so extreme. Those who do work come home exhausted and go immediately to bed so they can generate enough energy for work the next day. Because there isn't any apparent cause and no observable symptoms (like boils or measles), people with CFIDS are often confronted by people who just don't believe it's real.

COMMON SYMPTOMS OF CHRONIC FATIGUE SYNDROME[27]

Symptom	Percentage affected
Fatigue	100
Mental sluggishness, foggy thinking, inability to concentrate	50–85
Depression	50–85
Pharyngitis	50–75
Exhaustion after minimal exertion that can last for days	50–60
Stiffness and muscle pain	50–60
Muscle weakness	40–70
Joint pain, arthritis-like symptoms	40–50
Headache	35–85

In 1990 the Center for Disease Control in Atlanta began to keep records and study people with CFIDS in order to understand more about possible causes and therapies. We now know that CFIDS is multifactoral and affects many biochemical systems. There is an impairment in the ability of the mitochondria—the power plants in each of our cells—to produce energy. Cytokine production of interleukin 2 is low and causes poor immune function. Other immune parameters appear to be overstimulated. Although this seems paradoxical, it's probably not. According to Hans Selye, an expert on stress, our systems initially react to stress by overproducing. If working harder doesn't eventually solve the problem, they underproduce. Many people with CFIDS have exhausted adrenal glands and produce low amounts of cortisone and other adrenal hormones. Almost always they have dysbiosis, and most have Candida infections. There is usually leaky gut syndrome, accompanied by a host of food and environmental sensitivities. The liver is overburdened and overworked. The toxic by-products of life accumulate in tissues, and the cycle deepens.

There aren't any panaceas for CFIDS, but there are therapies that can gradually help restore people to health.

Restoration of digestive competency and nutrition go a long way towards normalizing CFIDS. Work with a nutritionally oriented health professional to design a program that meets your specific needs. The first steps are discovering any underlying problems which aggravate and drive the condition using the tests listed below. Then develop and follow a diet that is based upon foods that are healthful for you and a nutrient-rich program designed to boost immune, brain, and cellular function. When you are ready, add exercise—a little bit at a time.

The biological, rather than medical approach to chronic fatigue, saves money and works better. In one study of cost-effectiveness it was determined that a nutritional approach costs $2000 compared with $10,000 for a medical approach. The patients on nutritional programs reported greater improvements in function and subjective well-being. They were able to significantly reduce the amount of medications they used.[28]

The CFIDS Association in Charlotte, North Carolina, and the CFIDS Buyer's Club in Santa Barbara, California offer a wealth of information for people with CFIDS. You can obtain information about medications, herbs, nutritional supplements, diet, exercise and additional therapies. See listings in the Resources.

Functional Laboratory Testing

1. Comprehensive digestive and stool analysis with parasitology.
2. Elisa testing for food, airborne and chemical sensitivities. Elisa-Act test is the most inclusive.
3. Liver function profile.
4. Intestinal permeability screening.
5. Organic acids.

Healing Options

After testing, you'll have a better idea of any underlying problems. Look up related sections in this book to help you with the specifics. Then move on to the 4-R approach: Remove, Replace, Reinnoculate and Repair.

Metabolic cleansing: Metabolic cleansing involves going on a hypoallergenic food plan for one to three weeks and taking a nutrient-rich protein powder designed to help restore your liver's detoxification capacities. For a thorough discussion of metabolic cleansing, see Chapter 9.

Food and environmental sensitivities: Avoid all foods and chemicals that you are sensitive to from your diet for four to six months. Use shampoos, soaps and toiletries that are hypoallergenic for your specific needs and natural household cleaning products that are healthier for you, your family and the environment. Some people are sensitive to their mattresses, gas stoves, carpeting and upholstery. You may need to wear 100 percent cotton clothes and use 100 percent cotton sheets and blankets. Work with a health professional who can help you thread your way through the details.

Acidophilus and bifidobacteria: Supplemental use of beneficial bacteria can make a tremendous difference in your ability to digest foods. Beneficial flora can help reestablish the normal microbial balance in your intestinal tract. The supplements you purchase may have additional microbes, such as Saccharomyces. Dosage: 1 to 2 capsules or 1/4 to 1/2 tsp powder two to three times daily. Mix powdered supplement with a cool beverage and take on an empty stomach.

Digestive enzymes: Pancreatic or vegetable enzymes supply the enzymes that your body needs to digest fats, proteins and carbohydrates. Products differ. Some contain lactase, the milk digesting enzyme, others have additional hydrochloric acid to assist the stomach, and some contain ox bile to help with emulsification and digestion of fats. Dosage: 1 to 2 with meals.

Multivitamin with minerals: Because people with CFIDS have difficulty with absorption and utilization of nutrients, a highly absorbable, hypoallergenic nutritional supplement is necessary. Although products that contain herbs, bee pollen, spirulina and other additional food factors are good for many people, people with CFIDS often feel worse after taking food-based supplements. Make sure you buy the supplements that are herb and food free. Choose a supplement that contains the following nutrients: 25 to 50 mg zinc, 5000 to 10,000 IU vitamin A, 10,000 to 25,000 IU carotenes, 200 to 400 IU vitamin E, 200 mcg selenium, 200 mcg chromium, at least 25 mg of most B-complex vitamins, 400 to 800 mcg folic acid and 5 to 10 mg manganese.

Vitamin C: Vitamin C boosts immune function and helps detoxification pathways and has been shown to have antiviral effects. Clinicians have found it useful in people with CFIDS. Dosage: 3,000 to 5,000 mg daily. Do a vitamin C flush.

Magnesium: Found in green leafy vegetables and whole grains, magnesium is involved in over three-hundred enzymatic reactions in the body. It is essential for energy production, nerve conduction, muscle function and bone health.

People with CFIDS are often deficient in magnesium. Supplemental magnesium can improve energy levels and emotional states, while decreasing pain. Most people improve with use of oral magnesium supplements, but some need intravenous injections. Physicians can give 1000 mg magnesium sulfate by injection. In one study magnesium injections improved function in 12 out of 15 people, compared with only three receiving the placebo.[29] Dosage: 500 to 1000 mg magnesium citrate or magnesium/potassium aspartate (aspartic acid helps mobilize magnesium into the cells).

Co-Enzyme Q10: Co-Enzyme Q10 is necessary for energy production, immune function and repair and maintenance of tissues. It also enhances cell function. CoQ10 is widely used in Japan for heart disease and has been researched as an antitumor substance. Dosage: 60 to 100 mg daily. Trial period: 2 months: 90 to 100 mg daily.[30]

Essential fatty acids: Several studies have shown people with CFIDS to have fatty acid imbalances. In a recent study a combination of evening primrose oil and fish oil or a placebo of olive oil was given to 70 people with CFIDS. Of the people taking fish and evening primrose oils, seventy-four percent showed improvement at five weeks, and 85 percent showed improvement at 15 weeks. In comparison, the placebo group showed 23 percent improvement at 5 weeks and 17 percent at 15 weeks. Another study of the use of supplemental fatty acids showed improvement in 27 out of 29 people with CFIDS over 12 to 18 weeks. Twenty people who had previously been unable to work full-time for an average of over three years were able to go back to work full time after an average of 111 days. Sixteen months later 27 out of 28 who remained improved, and 20 were still progressing.[31] Dosage: Flax oil capsules 2 to 3 g daily or EPA/DHA fish oil once or twice daily plus evening primrose or borage oil 1 to 2 g daily.

Methionine: Methionine, an essential sulfur-containing amino acid, is commonly deficient in people with CFIDS. It acts as a methyl donor for transmethylation reactions throughout the body, especially in the brain. It also helps sulfoxidation for liver detoxification pathways and is a precursor for other sulfur-containing amino acids such as cysteine and taurine. People with CFIDS probably have an increased need for methionine. Some people find improvement with a general amino acid supplement that supplies methionine, lysine and carnitine simultaneously. Dosage: 500 to 1000 mg.

L-carnitine: People with CFIDS have low levels of acylcarnitine, an important amino acid.[1] Carnitine is vital for the conversion of fats into energy, plays some role in detoxification and is essential for heart function. Food sources of carnitine include meats, poultry and fish. Dosage: 500 to 1000 mg daily.

Lysine: Often people with CFIDS also have herpes infections. Some people find good results with a general amino acid supplement, which supplies carnitine, lysine and methionine as well as other amino acids. Dosage: 1 to 2 g daily at the first sign of an outbreak; 500 mg daily for prevention.

Malic acid: Malic acid comes from apples and is important in energy production at a cellular level. Several physicians have found malic acid supplementation reduces fatigue and pain of fibromyalgia.[32] Dosage: 6-12 tablets daily, decreasing dosage over time. Each tablet contains 300 mg malic acid/magnesium hydroxide.

Immune-stimulating herbs: Echinacea, goldenseal, astragalus, phytolacca (pokeweed), licorice and lomatium all have immune-stimulating properties. They can also help prevent secondary infections while you are in a susceptible state. Take them preventively or therapeutically as directed.

Stress management: Development of strong support systems is vital. People with CFIDS often have the illness for a long time and can greatly benefit from support groups. Exchange of information and dialogue with people who understand what you are going through can expedite recovery. Take time for yourself, rest and relax.

There is a new hypothesis that people with CFIDS are functioning in an anaerobic state so light anaerobic exercise may be most beneficial.[33] Working with light weights, leg lifts and use of weight machines to your capacity without causing fatigue may be more beneficial than aerobic exercise. As you begin to feel better, incorporate aerobic exercise—walking, biking, swimming and dancing. Prioritize so you have energy for what's most important. Be patient, kind and loving to yourself.

Eczema/Atopic Dermatitis

Eczema, or atopic dermatitis, is a catch-all name for any inflammation of the skin. It is a chronic skin condition characterized by redness, itching, and sometimes oozing, crusting, and scaling. The itch makes us scratch, which causes redness and inflammation. Approximately 1/2 to 1 percent of us have eczema and it affects 10 percent of all children. It can first appear at any age, but most often during the first year of life. Babies can have ezcema on their faces, scalp, bottom, hands, and feet. In children and adults, it may be more localized. The red patches are itchy, scaly, and dry, which encourages people to use lotions and creams to which they are often allergic. This complicates the problem further. Eczema varies over time, flaring up and calming down, at times better and worse. Emotional stress, heat, increase in humidity, bacterial skin infections, sweat, pets, dust, molds, pollens, toiletries, cosmetics, and wool clothing commonly aggravate eczema. As children age, they may continue to have eczema, it may disappear, or they may develop other allergies, including asthma. People with eczema have high levels of IgE, secretory IgA and eosinophils.

There are strong connections between eczema and food,

microbial, and inhalant allergies. Food allergies diagnosed through IgE and skin testing is apparent in one third of children with eczema.[34] Undoubtedly, many more would be borne out by IgG, IgA and IgM testing. Studies of children with eczema have shown that eczema improved in 49 out of 66 children after elimination of the particular foods. Foods that aggravate eczema, in descending order, were: eggs, cow's milk, food coloring, tomatoes, fish, goat's milk, cheese, chocolate and wheat. The longer one avoids these foods, the more likely the foods will cease to cause a problem. Elimination diets are recommended for children with eczema.

A study was made of 122 children, aged four months to six years, with food intolerance. Of them, fifty-two children had eczema; the rest had chronic diarrhea. The allergies caused damage to the intestinal lining, and there was a decrease in the body's ability to defend itself due to lactose intolerance and dysbiosis, which caused leaky gut syndrome, which led to more food antigens and sensitivity. Children with eczema had more damage than those with chronic diarrhea.[35]

Breast feeding dramatically reduces a baby's risk of developing eczema and allergies in general. Babies with eczema, and probably most babies, should not be given cow's milk, milk products, eggs or wheat before one year of age. As their digestive system matures, they can better handle these complex foods. Babies with eczema who drink formula should be tested by skin prick to determine which formulas are most suitable for them.

Elimination of foods, stress and allergens can significantly alter the course of the disease. Even though you cannot control all factors, controlling enough of them will allow you to stay under the symptom threshold. Jonathan Wright, M.D. had success in 39 out of 40 patients with eczema who followed this program: 50 mg zinc three times daily for six weeks, plus 2 mg copper daily, 5 g Omega 3 fatty acids (evening primrose or borage oil) twice a day for three months and Omega 6 fatty acids (EPA/DHA fish oils) 1 to

2 g three times daily for four weeks. If you decide to use his protocol, work with a health professional.

The most common treatment for eczema is cortisone cream, which suppresses your body's normal immune function. However, there is often a rebound effect after you stop using the cream, and your symptoms return worse than before. Natural creams with chamomile, licorice and comfrey root are very effective at soothing and healing eczema without negative side effects.

Functional Laboratory Testing

1. Allergy testing for IgE, food, mold, dust and inhalants.
2. Sensitivity testing, Elisa testing.
3. Complete digestive and stool analysis. Check to see if there is dysbiosis and how well the body is digesting. Some studies indicate a low level of hydrochloric acid in people with eczema, and the CDSA can give a general indication of HCl production.
4. Intestinal permeability screening. It will most likely be positive.
5. Heidelberg testing for adequacy of hydrochloric acid production.

Healing Options

Food and environmental sensitivities: Follow the directions in Chapter 9. An elimination/provocation diet (discussed in Chapter 7) can significantly reduce eczema. Foods that you are sensitive to will generally make you itch. Often the itching starts soon after the meal, but it can be delayed up to 48 hours, which makes tracking the foods down a bit tricky. Food allergy and sensitivity testing can help you determine which foods to eliminate from your diet.

Avoidance: Eliminate all foods and chemicals that you are sensitive to for four to six months. Use natural household cleaning products and shampoos, soaps and toiletries that are hypoallergenic. If you are sensitive to mattresses, gas stoves, carpeting and upholstery, you may need to use cotton clothing and sheets that allow the skin to breathe naturally. Work with a health professional who knows how to help you meet your needs.

Lactobacillus and bifidobacteria: Restoring the normal balance of flora in your intestinal tract can help reduce eczema. Use of beneficial bacteria supplementally can make a tremendous difference in your ability to thoroughly digest foods. The supplement you purchase may have additional microbes, such as Saccharomyces. Dosage: 1 to 2 capsules two to three times daily, or 1/4 to 1/2 tsp powder two to three times daily. Mix powdered supplement with a cool beverage. It works best on an empty stomach.

Babies and small children with eczema: Babies are born with sterile digestive tracts, and as soon as they are born they are exposed to microbes of all sorts. Dairy and soy products are difficult to digest until the baby's digestive system has mature flora. Supplementing with beneficial flora, *Bifidobacteria infantum* and others will help your baby digest food more easily and heal the eczema. Use probiotic formulas specifically designed for infants, toddlers or small children. Dosage: 1/8 to 1/4 tsp three times daily. If the mother is breast feeding, she should also take 1/4 tsp three times daily.

Check for Candida infection: Fungal infections are a common cause of eczema. In a study of 115 men and women with eczema, 85 were sensitive to fungus and after they were treated with fungal creams, oral ketakonazole or a yeast-free diet, there was much improvement.[36] Take the yeast self-test and do blood testing or complete digestive and stool analysis to determine if yeast is contributing to your eczema.

Natural eczema creams: Herbal creams can be as effective as cortisone creams in reducing eczema, and they don't have the negative side effects. Chamomile creams are widely used in Europe. Licorice root stimulates production of healing and anti-inflammatory prostaglandins. Comfrey root contains allantoin which promotes healing. Some products contain all three of these ingredients. Look in health food stores or ask your health professional to find a product that works for you.

Multivitamin with minerals: It is wise to add a good quality multivitamin with minerals to your daily routine. However, you probably cannot get all the nutrients you require from a single tablet or capsule. Look for a supplement that has *at least* 100 mcg chromium, 100 mg selenium, 5 to 10 mg manganese, 500 mg calcium, 250 mg magnesium, 25 to 50 mg zinc, 1 to 2 mg copper, 10,000 to 25,000 IU vitamin A (not beta carotene; pregnant women should not exceed 10,000 IU daily) and 400 IU vitamin E. Zinc and vitamin A are essential for healthy skin and mucous membranes, and zinc is also necessary for production of anti-inflammatory prostaglandins and formation of hydrochloric acid. There are many anecdotal tales about the effectiveness of vitamin E and eczema, but no controlled studies.

It is best *not* to buy a supplement that contains many different foods and herbs because you may be unknowingly sensitive to one or more of the ingredients. Be sure to buy a supplement that is free of common allergens.

Quercetin: Quercetin, the most effective bioflavonoid for anti-inflammation, can be used to reduce pain and inflammatory responses and control allergies. Dosage: 500 to 1000 mg three to four times daily.

Vitamin C: A study of ten young people with severe eczema showed that supplementation with vitamin C significantly improved eczema and immune function. They needed only half as many antibiotics for treating skin infections as the control group.[37] Dosage: 1000 to 3000 mg mineral ascorbates or Ester-C daily. Do a vitamin C flush once a week.

Neutralizing reactions: There are many ways to minimize the effects of food sensitivities. Clinical ecologists can provide neutralization drops to counteract your reaction to particular foods. These drops work like allergy shots—a small amount of what you are sensitive to helps stimulate your body's natural immune response. Malic acid can also curtail sensitivity reactions.

Evening primrose oil and flax seed oil: Studies show that people with eczema generally have low levels of both Omega 3 and Omega 6 fatty acids. The first step in metabolism of linoleic acid, which allows for the conversion into gamma-linolenic acid (GLA), is often impaired in people with eczema. Taking GLA directly in evening primrose, flax or borage oil circumvents blockage.[38] GLA has an anti-inflammatory effect and benefits immune function. Dosage: 1 to 2 gram three times daily. Trial period: four weeks.

Fish Oils: One recent study on people with eczema showed a 30 percent improvement in a four-month trial of 8 capsules of fish oil per day. Though the placebo group was given corn oil, which gave an improvement of 24 percent, results suggest that people with eczema have a generalized need for essential fatty acids.[39]

Eating cold water fish—salmon, halibut, sardines, herring, tuna—two to four times each week can provide you with the Omega 3 oils you need. If you use fish oil capsules, do so under the supervision of a physician. They cause a significant increase in clotting time and should not be used by people with hemophilia or those on aspirin or anti-coagulant drugs.

Nickel-restricted diet: The relationship between nickel sensitivity and eczema has appeared recently in scientific literature. Nickel is an essential nutrient that is found in many enzymes. However, excess nickel is an irritant to the gastrointestinal lining. You can be tested for nickel sensitivity through skin testing or an oral challenge. Nickel is used as an alloy in jewelry, so if jewelry irritates your skin or turns it gray, you may be sensitive to nickel. If you are, it is recommended that a low-nickel diet be followed for a limited period of time. High-nickel foods are chocolate, nuts, dried beans and peas, and grains.

Fibromyalgia and Myofascial Pain Syndrome

Fibromyalgia is characterized by long-term muscle pain and stiffness. It used to be called "fibrositis," which implies an inflammation of fibrous and connective tissues like muscles, tendons, fascia and ligaments. Myofascial pain syndrome is characterized by just a few painful and achy places, most often in the jaw, that are tender when trigger points are touched. Fibrositis is a nonspecific term that identifies both fibromyalgia and myofascial pain syndrome.

Fibromyalgia is characterized by generalized aching, pain and tenderness throughout the body. People complain of neck, shoulder, low back and hip pain that seems to move around from place to place. People often report changes in sleep patterns and wake up often during the night with a feeling of achiness, stiffness and fatigue. Other symptoms occur less frequently and include intolerance to cold or heat, bladder problems, Raynaud's phenomenon, nasal congestion, teeth grinding, abdominal pain, constipation or diarrhea, headaches, numbness or swelling in hands and/or feet, anxiety and depression. People with fibromyalgia often report a traumatic event that triggered initial symptoms: emotional or physical stress, an accident or a severe infectious illness. Though common, this doesn't occur in everyone.

Fibromyalgia occurs more in women than in men and has an incidence of 1 percent in people under 60 and 4 percent in people over 60. Fibromyalgia shares many symptoms with chronic fatigue syndrome, though it is classified as its own disease. Recent studies indicate that myofascial pain, fibromyalgia and CFIDS are on a continuum of the same disease path, with myofascial pain being the mildest, fibromyalgia moderate and CFIDS the most severe.[40]

People with fibromyalgia are often misdiagnosed and go from doctor to doctor in search of help. They are generally put on anti-inflammatory drugs. One study showed that 90 percent of people treated this way were still symptomatic after three years.[41] Leo Galland, M.D. finds food and environmental sensitivities, candida or parasites to be causal factors in fibromyalgia.[42] When the underlying problem has been identified and treated, fibromyalgia resolves.

A very small, but promising study was done with three people who had fibromyalgia for five to ten years. They were tested for food and environmental sensitivities with the Elisa/Act test and given dietary restrictions. They were put on a detoxification program and personalized nutritional therapies to meet their needs and stimulate repair of cells and tissues. The final component was stress management, with recommendations for relaxation training, exercise and biofeedback. In six to twelve weeks these people showed a reduction of 80 to 90 percent in their symptoms. They also showed a significant reduction in the number of foods and environmental sensitivities in repeated Elisa testing.[43] More research needs to be done in this area.

Nutritional therapies have been successful in the reduction of symptoms. Use of a single supplement may bring some relief, but a total program is necessary to bring dramatic relief and true healing. Taking Co-enzyme Q10, vitamins B1, B6 and C, arginine, tryptophan, essential fatty acids, antioxidants, niacin and magnesium malate (magnesium plus malic acid) in addition to a hypoallergenic diet has been shown to have positive effects.[44]

FUNCTIONAL LABORATORY TESTING

1. Liver function profile.
2. Elisa food and environmental sensitivity testing.
3. Intestinal permeability screening.
4. Oxidative stress evaluation.
5. Comprehensive digestive and stool analysis.

HEALING OPTIONS

Metabolic cleansing: Metabolic cleansing involves going on a hypoallergenic food plan taking a nutrient rich protein powder designed to help restore your liver's detoxification capacities. Use this protocol for one to three weeks. See Chapter 9 on detoxification and metabolic cleansing.

Food and environmental sensitivities: Eliminate all foods and chemicals that you are sensitive to for four to six months (see discussion in Chapter 7). Work with a health professional who can help you find your way through the details.

Acidophilus and bifidobacteria: Use of beneficial bacteria taken supplementally can help reestablish the normal microbial balance in your intestinal tract. The supplement you purchase may have additional microbes as well. Dosage: 1 to 2 capsules two to three times daily or 1/4 to 1/2 tsp powder two to three times daily. Mix powdered supplement with a cool beverage and take on an empty stomach.

Digestive enzymes: Pancreatic or vegetable enzymes supply the enzymes your body needs to digest fats, proteins and carbohydrates. Some products contain lactase, the milk-digesting enzyme, others have additional hydrochloric acid to assist the stomach, and some contain ox bile to help with emulsification and digestion of fats. Dosage: 1 to 2 with meals.

Multivitamin with minerals: A high-quality hypoallergic nutritional supplement is necessary. Although products that contain herbs, bee pollen, spirulina and other additional food factors are good for many people, it's best to buy supplements that are herb- and food-free. Look for the following levels of specific nutrients: 50 to 100 mg vitamin B1, 50 to 100 mg vitamin B6, 200 to 400 IU vitamin E, 10,000 IU vitamin A, 10,000 to 25,000 IU carotenes, 200 mcg selenium, 200 mcg chromium, 5 to 10 mg manganese, glutathione, cysteine or N-acetyl cysteine, plus additional nutrients. Antioxidant nutrients—carotenes, vitamins C and E, selenium, glutathione, Co-Q-10, cysteine and NAC—have been shown to be needed in larger quantities in people with fibromyalgia. In a recent study people with fibromyalgia had an increased pyruvate-to-lactate ratio, which may indicate the need for more vitamin B1 and B6.[45]

Vitamin B1: People with fibromyalgia have lower levels of red blood cell transketolase, which is a functional test for vitamin B1 (thiamin) status. Researchers found that supplemental thiamin pyrophosphate worked better than other forms. This suggests a metabolic defect rather than a true deficiency. This may also reflect a magnesium deficiency because thiamin-dependent enzymes require magnesium.

Vitamin C: Vitamin C boosts immune function, helps detoxification pathways and has anti-viral effects. Dosage: 3,000 mg daily. Once a week, do a vitamin C flush. (See directions for C Flush, pages 147–148.)

Magnesium: It is very common for people with fibromyalgia to be deficient in magnesium. Supplemental magnesium can improve energy levels and emotional states while decreasing pain. Most people improve by using oral magnesium supplements, but some need an intravenous injection of 1000 mg magnesium sulfate by a physician (for more on magnesium, see discussion under Chronic Fatigue Syndrome). Choline citrate can greatly enhance oral magnesium utilization (available from Perque/Seraphim listed in Directory). Dosage: 500 to 1000 mg magnesium citrate or magnesium/potassium aspartate.

Malic acid: Malic acid found in apples is important in energy production at a cellular level. Several physicians have found malic acid supplementation reduces fatigue and the pain of fibromyalgia. Dosage: 6 to 12 tablets of 300 mg malic acid/magnesium hydroxide daily, decreasing dosage over time.

Co-enzyme Q10: Co-Enzyme Q10 is necessary for energy production, immune function, repair and maintenance of tissues and enhanced cell function. Dosage: 60 to 100 mg daily. Trial period: two months.

Quercetin: Quercetin is the most effective bioflavonoid in its anti-inflammatory effects and can be used to reduce pain and inflammatory responses and control allergies. (For more on quercetin, see discussion under Osteoarthritis.) Dosage: 500 to 1000 mg three to four times daily.

Glucosamine sulfate: Glycosamine sulfate is used therapeutically to help repair cartilage, reduce swelling and inflammation, and restore joint function, with no reported side effects. Dosage: 500 mg two to four times daily.

Gamma oryzanol: Although no specific studies have been done with fibromyalgia, gamma oryzanol, a compound found in rice bran oil, heals damage to the intestinal tract and normalizes serum triglycerides, cholesterol, symptoms of menopause and depressive disorders.[46] Dosage: 100 mg three times daily for a trial period of three to six weeks.

Glutamine: Glutamine, the most abundant amino acid in our bodies, is used in the digestive tract as a fuel source and for healing stomach ulcers, irritable bowel syndrome, ulcerative bowel diseases and leaky gut syndrome. Dosage: Begin with 8 g daily. Trial period: four weeks.

Arginine: People with fibromyalgia have been shown to have lower levels of arginine than other people. Dosage: 500 to 1000 mg or a mixed amino acid.

Tryptophan: People with fibromyalgia have lower tryptophan levels than controls.[47] Tryptophan is a precursor to serotonin, a neurotransmitter that helps us sleep and prevents depression. Unfortunately, tryptophan is only available by perscription at a few pharmacies nationwide due to a contaminated batch several years ago. Passion flower, an herb with high levels of tryptophan, has been used historically for depression, anxiety and insomnia, all of which are symptoms of fibromyalgia. Tryptophan is also found in cashews, cheddar cheese, eggs, halibut, peanuts, salmon, sardines, shrimp, turkey and tuna.

Cayenne/Capsaicin: The prescription drug Capsaicin, a cayenne pepper cream, was used in a study of 45 people with fibromyalgia. It was found to improve grip strength and reduce pain over a two-week period.[48] Capsaicin cream burns temporarily, but this diminishes over time.

S-adenosyl-L-methionine: A recent study of 47 people with fibromyalgia showed that injections and oral supplementation of S-adenosyl-L-methionine significantly reduced muscle tenderness and the number of tender points, lowered pain severity and benefited depression and anxiety.[49] S-adenosyl-L-methionine is produced in our bodies from methionine. It is the active methylating agent for many enzyme reactions throughout the body, especially in the brain. People with fibromyalgia can probably make this conversion, so oral methionine may be useful clinically.

Electro-acupuncture: In a study of electro-acupuncture with people with fibromyalgia, half the people received acupuncture on four points; the other half received acupuncture 20 mm away from the actual point. Seven out of eight symptom parameters improved in the four-point electro-acupuncture treated group. None changed in the placebo group. In the electro-acupuncture group, a quarter had no improvement, half improved satisfactorily and a quarter had their symptoms disappear completely.[50] Although this study was done with electro-acupuncture, similar results would occur with classical acupuncture.

FOOD AND ENVIRONMENTAL SENSITIVITIES

Food sensitivities are the result of leaky gut syndrome. It's important to discover the underlying cause, which may be parasites, candida infection, bacterial or viral infection, pancreatic insufficiency, medications or poor lifestyle habits. Two types of allergies cause reactions: acute or immediate hypersensitivity reactions (Type I) and delayed hypersensitivity reactions (Type II-IV).

True food allergies (Type 1) are rare. They affect from 0.3 to 7.5 percent of children and 1 to 2 percent of adults. Type IgE antibodies bind to the offending food antigens and cause the release of cytokines and histamines which results in hives, skin rashes, closing of the throat, respiratory distress, runny nose, itching and sometimes severe reactions of asthma and anaphylactic shock. These symptoms occur within minutes after the food is eaten. Foods which most often trigger these reactions are eggs, cow's milk, nuts, wheat, soy, white fish and shellfish. Physicians diagnose food allergies through the use of patch skin tests and RAST blood testing, but these tests do not accurately determine food or environmental sensitivities.

Type II-IV reactions are the result of food or environmental factors and cause symptoms which are delayed, taking several hours to several days to appear. This makes tracking them down very difficult. They are more common than true food allergies, affecting 24 percent of American adults.[51] Food sensitivities cause a wide number of symptoms resulting from a typical leaky gut reaction. (See chart listing common symptoms of food and environmental sensitivities in Chapter 6.) Food particles enter the bloodstream through damaged mucosal membranes, the body recognizes them as foreign substances (antigens) and triggers an immune reaction to get rid of them. Prolonged antibody response overwhelms the liver's ability to eliminate the antigens, so the toxins enter the bloodstream and trigger delayed hypersensitivity response, inflammation, cell damage and disease. Almost any food can cause a reaction, although the foods that provoke 80 percent of food sensitivity reactions are wheat, beef, dairy products, egg, pork and citrus.

Two types of food sensitivities are recognized by doctors of preventive medicine and clinical ecology: cyclic and fixed. Cyclic account for 80 to 90 percent of food sensitivities. These reactions occur after a specific food has been eaten over and over, causing a reaction. If you avoid eating the food for four to six months, your body will most likely tolerate it again. However, you must correct the underlying leaky gut syndrome, or the problem will reoccur with that or other foods. With the fixed type, sensitivities don't go away, even if you have avoided eating the foods for periods of time. Occasionally, you can eat it after years of avoidance without provoking symptoms.

These reactions happen when antibodies are triggered in response to foods, chemicals, and bacterial toxins. They damage cells and inflammation occurs due to the damage. Many people are aware of which foods and chemicals bother them, but just as often the cause is hidden. With repeated exposure to these foods our bodies slowly adapt to the irritants and the symptoms that are provoked.

Usually we learn to live with the symptoms. Remember your first cigarette or glass of beer? It was probably distasteful to you. You probably didn't like the feeling of the smoke in your throat and lungs, and didn't like the taste of beer. With continual use of beer or cigarettes, those initial reactions disappear, and eventually we like and even crave them. They still have a negative effect on our system, but our body has adapted to it. This model works for our relation to foods and chemicals as well. Habitual use dulls our ability to recognize their negative health effects. Because of this phenomenon, we can make use of the elimination provocation challenge.

The idea behind the elimination provocation challenge is simple—only eat foods that you are unlikely to be sensitive to for a week or two, and then add back the foods you normally eat to "challenge" your system. Removal of offending foods calms down symptoms. Challenging yourself with these foods allows you to determine which foods trigger symptoms. Careful addition of only one food each two days makes it easier to determine which foods caused the reaction.

While the elimination provocation challenge sounds simple the administration of it can be tricky. In general, people have no problem with the elimination part—a restricted food plan for a week can be accomplished with a bit of planning and discipline. Adding foods back into your diet slowly is more difficult. Symptoms are delayed and recipes contain a combination of foods, so that you can't easily tell which ones cause the reaction. Eating simple, or single foods is helpful for determining which foods you may be sensitive to. You may feel badly after eating something, but not be certain which ingredient caused the distress. It then becomes necessary to remove all suspected foods for four days, and try them again one at a time. If you have the same reaction each time you add the food, you've found the culprit. If you have a food sensitivity to one food, you are often sensitive to all foods in the same food family. For example, some people who are sensitive to wheat are sensitive to all grains in the grass family. People who are sensitive to milk are often sensitive to cheese and whey. It is common to be sensitive to more than one food or food family.

Environmental illness, also called multiple chemical sensitivities, is becoming more and more common. In fact, two subspecialties of medicine are now devoted to it: environmental medicine and clinical ecology. Chronic exposure to food additives, household chemicals, building materials, recirculating air and impure water can so depress and weaken our immune systems that even a small amount of toxic exposure makes us ill.

To test for food allergies, run a RAST or modified RAST test, which checks for elevation of IgE antibodies. Elisa tests for food and environmental sensitivities check levels of IgG4 and sometimes IgM and IgA antibodies. You can also do an elimination/provocation test at home. Though highly accurate, it can be frustrating. I prefer to use it along with a blood test, which helps reinforce and simplify the program.

If you have cyclic food or environmental sensitivities, you'll do best with a holistic approach. By avoiding substances you're sensitive to for a period of six months, your body will gradually stop reacting to most of them. It really helps the process if you also detoxify the body and the liver (see Chapter 9). Natural foods, organically grown and nutrient rich, also help repair the body.

For people who are highly sensitive to foods, a four-day food rotation may be helpful in addition to complete avoidance of foods you are highly sensitive to. The theory behind the rotation diet is that when you eat a food, your body begins to produce antibodies against it. If you don't eat it again for several days, the antibodies are no longer present. When you eat the food on day five, the process begins again, but you never develop symptoms because the antibodies are never prepared at the correct moment. There are many good books on the four-day food rotation diet. My favorite is *If It's Tuesday, It Must Be Chicken* by Natalie Golos and Frances Golos Golbitrz.[52]

To set up the rotation, make four lists, dividing all foods into four groups, with a quarter of your fruits, vegetables, grains, proteins, oils, nuts, seeds and beverages in each group. Keep all foods from the same family in the same group. For example, all grains are in one family because of the similarity of their makeup. Follow these food lists in order and then begin again with the first list on the fifth day.

A comprehensive program of nutritional supplements will also help the cells regenerate and generally aid the healing process. As noted in the Elisa/Act patient handbook, "Persons suffering from immune system dysfunction and overload due to delayed hypersensitivity reactions often have a need for even greater supplementation because of poor functioning of the body's normal biochemical pathways."[53] Exercise programs and stress management tools also play a part in recovery. If you follow a holistic program, you'll find that over time you will become less and less sensitive to foods and the environment.

Functional Laboratory Testing

1. Elisa Testing—IgG4 and possibly IgM and IgA.
2. RAST or modified RAST for IgE.
3. Candida testing, either serum or through CDSA.
4. Comprehensive digestive and stool analysis with parasite screening.
5. Intestinal permeability test (almost surely will be positive).

Healing Options

Avoidance: Eliminate all foods and chemicals that you are sensitive to for four to six months. Health food stores specialize in wheat-free breads, rice noodles and a plethora of foods for people with food allergies. If you are sensitive to chemicals, use natural household cleaning products. Some people are sensitive to mattresses, gas stoves, carpeting and upholstery, which can make elimination tricky. Work with a health professional who can help you with the details.

Neutralizing reactions: There are many ways to help minimize the effects of food sensitivities. Clinical ecologists can provide neutralization drops to desensitize you to reactive foods. These drops work like allergy shots—a small amount of what you are sensitive to stimulates your body's natural immune response. Malic acid is also useful to neutralize sensitivity reactions.

Quercetin: Quercetin, the most effective bioflavonoid thanks to its anti-inflammatory effects, can be used to reduce pain and inflammatory responses and control allergies. Dosage: 500-1000 mg three to four times daily.

Tri-salts: Alkalizing mineral salts that generally contain calcium, magnesium and/or potassium bicarbonates help minimize reactions to foods. Alka-Seltzer Gold can be used in this capacity, since it contains sodium and potassium salts.

Gamma-oryzanol: Gamma-oryzanol, found in rice bran oil and products that help heal leaky gut syndrome, has been shown to be effective in normalizing serum triglycerides and cholesterol, symptoms of menopause and depressive disorders. Dosage: 100 mg three times daily. Trial period: three to six weeks.

Glutamine: The most abundant amino acid in our bodies, glutamine is effective for healing the intestinal tract, including stomach ulcers, irritable bowel syndrome and ulcerative bowel diseases. Dosage: 8 g daily. Trial period: four weeks.

Digestive enzymes: If the pancreas cannot make or recirculate enough digestive enzymes, we don't digest foods sufficiently. Use of either vegetable or pancreatic enzymes helps you break down foods more completely. Dosage: 1-2 capsules/tablets with meals.

Multivitamin with minerals: To cure problems with leaky gut, it is wise to add a good quality, allergen-free multivitamin with minerals to a daily routine. Look for a supplement that has *at least* 100 mcg chromium, 100 mg selenium, 5-10 mg manganese, 500 mg calcium, 250 mg magnesium, 15 mg zinc. It is best to buy a supplement that does not contain a lot of foods and herbs because you may be sensitive to one or more of the ingredients in the supplement itself.

Vitamin C: Vitamin C helps flush toxins from our bodies, so take extra vitamin C. Dosage: 1000-3000 mg mineral ascorbates or Ester-C daily. Vitamin C flush once every week or two.

MIGRAINE HEADACHES

Migraine headaches cause periodic disruption in the lives of 11 million Americans, affecting 6 percent of men and 15 to 17 percent of women every year.[54] Migraines occur when blood vessels in the head contract and then dilate and swell.

Migraines usually begin with a throbbing pain on one side of the head, which can spread to both sides. They are often preceded by a short period of mood changes, irritability, fatigue, numbness or tingling on one side of the body, lack of appetite, changes in vision or seeing bright spots. These symptoms may disappear when the headache appears or remain. Although symptoms vary from person to person, they have a consistent pattern in each individual. Migraine attacks may last from hours to days and may be accompanied by nausea, vomiting and extreme sensitivity to light. Medications and other techniques work best if used at the beginning stages.

Migraines usually come on in response to a "trigger." Common triggers are foods and beverages, alcohol, stress, emotions, hormone changes, medications such as estrogen therapy, visual stimuli or changes in routine. A recent study of 494 people with migraines cited the following triggers: stress in 62 percent, weather changes in 43 percent, missing a meal in 40 percent and bright sunlight in 38 percent. Cigarettes, perfumes and sexual activity also provoked migraines in some people.[55] Other triggers are red wines, exhaustion and monosodium glutamate (MSG).

Jean Munro, MD, an English doctor who specializes in working with people with multiple chemical sensitivities, breaks migraines into four types. The first type is a classic migraine, which begins with a visual disturbance of some sort—flashing lights, blackening or blurred vision. It usually involves one side of the head, and people often vomit. The migraine usually lasts one to three days and can be quite severe.

The second type is called a common migraine and is almost identical to the first except that there is no visual warning. It begins on one side, sometimes progressing to both, and there may be vomiting. The third type is called a basilar migraine, when the blood vessels at the base of the head dilate. It can be quite frightening and often causes a panicky feeling, accompanied by a sense of doom. A generalized headache is accompanied by a pins-and-needles sensation around the mouth, nausea and tingling hands. The fourth type, called a motor migraine, is a variation on the basilar and may be quite severe. Half the body feels weak, head pain centers around the eye, and vision is distorted.

Of 282 patients with migraines whom Dr. Munro studied, 100 percent had food allergies or sensitivities. Over 200 of them were sensitive to wheat and/or dairy products. Other commom trigger foods were tea, oranges, apples, onions, pork and beef. She found that foods eaten daily provoked more reactions than chocolate, alcohol and cheese, which are thought to be the most common triggers. Dr. Munro also found that people who eliminated these foods from their diet and cleared their homes of environmental contaminants had the best results in prevention of migraines. Using mild household cleaners, getting rid of gas appliances, removing house plants with molds and fungus, frequent cleaning and making a bedroom an oasis by removal of carpets and curtains resulted in fewer migraines.[56] Although these people were still exposed to smoke, perfume and other environmental triggers outside the home, changing the home environment and their diets lowered their threshold enough so that they became more tolerant. More recent studies show that IgG4 and anti-IgG antibodies increased after food challenges in people with migraines, which supports the food-sensitivity/allergy hypothesis.[57] Other researchers have confirmed that many of the foods that Dr. Munro found also provoked symptoms. However, virtually any food can be a trigger.

John Diamond, MD, has used the theory in his Diamond Headache Clinic in Chicago that foods high in amines also provoke migraines in some people. Dietary amines, which promote constriction of blood vessels, are normally broken down by enzymes, but some people with migraines have lower than normal amounts of the appropriate enzymes. The amines that provoke vaso-constriction are serotonin, tyramine, tryptamine and dopamine. They are found in the greatest quantities in avocados, bananas, cabbage, eggplant, pineapple, plums, potatoes, tomatoes, cheese, canned fish, wine (especially red), beer, aged meats and yeast extracts.[58]

Hormone fluctuations in women can worsen, improve or trigger migraines. Many women only experience migraines at specific times in their menstrual cycle from ovulation through menstruation. Birth control pills and other estrogen-containing medications are widely recognized to trigger migraines in susceptible women. When women stop taking the medications, their migraines typically disappear.

The truth is that migraines have many triggers that vary from person to person. Finding your triggers and the treatments that work best for you is key. You certainly won't need all the therapies listed below, but hopefully you'll find relief from some of them.

Functional Laboratory Testing

1. IgG4 and IgE Elisa testing for food allergies/sensitivities.
2. Elisa testing for environmental sensitivities/allergies.
3. Intestinal permeability screening.
4. Comprehensive digestion and stool analysis with parasitology.
5. Candida antibodies or CDSA.

Healing Options

Food sensitivities and allergies: Avoid foods you are sensitive to. Make your home environmentally safe by using only natural cleaning supplies, removing gas appliances, cleaning out mold and mildew, using a dehumidifier and making your bedroom into a safe harbor by removing unnecessary items, carpeting and drapery.

Candida: A recent study of the relationship between candida and migraines found that 13 out of 17 migraine sufferers responded to a three-month program of diet and medication with fewer and less severe headaches. Blood testing showed a lowering of candida antibodies as well. The four people who did not respond well didn't stick to the program![59] If you have migraines, take the home test for candida infections and have your physician order additional testing.

Eat often: Low blood sugar levels often trigger migraines so don't skip meals. You may find that eating five to six small meals each day works better for you than three main meals.

Magnesium: Numerous studies have documented the relationship between low magnesium levels and migraine headaches.[60] When magnesium is supplemented at levels of 600 mg daily, there is a significant decrease in the number of migraines. Alan Gaby, M.D., offers his patients magnesium injections (of B-complex, vitamin C and calcium) which alleviate their migraine headaches within minutes.

Feverfew: Numerous studies have shown the herb feverfew (*Tanacetum parthenium*) to be effective in preventing and minimizing the severity of migraines. The active ingredient is believed to be sequiterpene lactone. Feverfew affects arachidonic acid metabolism, which inhibits the generation of leukotrienes and thromboxane that cause inflammation and pain. Feverfew also inhibits secretion of serotonin from platelets and has a dampening effect on substances that cause blood vessels to constrict. One classic study took people who had benefited from feverfew and put them on feverfew or a placebo. The people on the placebo had a significant increase in the number and intensity of headaches.[61]

Feverfew needs to be taken on a daily basis as a prevention rather than as a medication. There is a difference between fresh and dried feverfew and between various samples. If you don't get relief from one type, try another. Fresh feverfew seems to work best. It is easy to grow, so you could just eat a few leaves each day. Tinctures are available and would best approximate fresh leaves. It also comes in a freeze-dried form that seems to be effective. Dosage: Twice daily 15-20 drops tincture, 1 capsule or 1 to 3 fresh leaves.

Riboflavin (vitamin B2): Forty-nine people with recurrent migraines were given 400 mg of vitamin B2 daily with breakfast for three months. The number of migraines declined by 67 percent and the severity diminished by 68 percent.[62]

Acupuncture: Acupuncture has been shown to reduce the incidence and severity of migraine headaches along with the drug metoprolol in controlled studies. Metoprolol was slightly more effective, but acupuncture produced fewer negative side effects.[63]

Behavioral techniques: Biofeedback, hypnotherapy and stress reduction techniques have all proven useful to some migraine sufferers. Behavioral techniques help us better understand stressors and how to cope more effectively.

Avoid monosodium glutamate: MSG can provoke migraine headaches, asthma, diarrhea, vomiting and gastric symptoms. These problems can occur immediately after eating or may be delayed up to 72 hours, which makes their relationship to MSG more difficult to discover. Food product labels may be misleading, with MSG labeled "natural coloring"; some hydrolyzed vegetable protein contains MSG. You can challenge yourself with MSG to see if it brings on a migraine. The Elisa-Act blood test includes tests for MSG and glutamate sensitivity.

Quercetin: Quercetin, the most effective bioflavonoid due to its anti-inflammatory effects, can be used to reduce pain and inflammatory responses and control allergies. Dosage: 500-1000 mg three to four times daily at onset of migraine; 500 mg daily as a preventive.

Omega 3 fatty acids: Fish oils have been shown to have a variable effect with migraines. Some people find complete prevention, others experience no change, and others get worse. Depending on your reaction, it may be beneficial to eat fish that contain high amounts of EPA/DHA oils—tuna, mackerel, halibut, salmon, herring and sardines—two to four times a week. Dosage: 1-2 g capsules daily. (Higher doses have been used in studies, but you should only increase dosage under the supervision of a physician.)

Chiropractic and massage: Chiropractic manipulation and massage can help blood and lymphatic supply and lessen muscle tension.

Caffeine: Caffeine plays a mixed role in migraines. For some people it significantly reduces the number and severity of headaches; for others it triggers them.

Niacin: There have been reports of physicians who use niacin intravenously during a migraine to lessen severity and duration of headaches. The dose contains at least 100 mg niacin, which is infused slowly.[64]

Antioxidants: Migraines are often triggered by substances that promote free radicals like cigarette smoke, perfume, hair spray, pollution and household chemicals. One researcher found lower levels of superoxide dismutase in platelets of people with migraines than in people with tension headaches.[65] More research needs to be done in this area, but taking adequate antioxidants in a multivitamin with minerals may help prevent migraines.

Psoriasis

Psoriasis is a chronic skin rash characterized by scaling, patchy, silvery-looking skin. It can affect just knees, elbows or scalp or can spread over most of the body. It often occurs at the site of a previous injury. Psoriasis often runs in families and usually develops gradually. It's believed that the thick scaling of psoriasis is due to overproduction of new skin and a rapid rate of skin breakdown. Psoriasis flares up due to stress, severe sunburn, irritation, skin creams, antimalarial therapy or from withdrawal from cortisone, or it can be brought on by other triggers. Then it goes into remission. Psoriasis affects about 1 percent of the American population as a whole, but 2 to 4 percent of Caucasians.

Psoriasis can also occur with joint inflammation as psoriatic arthritis (see discussion on arthritis) and is found in 5 to 7 percent of people with psoriasis. It isn't clear whether psoriasis and psoriatic arthritis are the same disease or two almost identical diseases.

Michael Murray, N.D. and Joe Pizzorno, N.D. have documented a number of factors that influence the progression of psoriasis, including incomplete digestion of protein, bowel toxemia, food sensitivities, poor liver function, reaction to alcoholic beverages and eating high amounts of animal fats.[66] Let's look at each of these factors.

When protein digestion is incomplete or proteins are poorly absorbed, bacteria can break down the proteins and produce toxic substances. One group of these toxins is called polyamines, which have been found to be higher in people with psoriasis than in the average population. Polyamines contribute to psoriasis by blocking production of cyclic-AMP. Vitamin A and an herb called goldenseal inhibit the formation of polyamines. Since protein digestion begins in the stomach, low levels of hydrochloric acid there can also cause incomplete protein digestion. Digestive enzymes and/or hydrochloric acid supplementation aid protein digestion.

Bowel toxemia plays an important role in psoriasis. A poor balance of intestinal flora due to stress, diet, medications or other factors often leads to bacterial and fungal infection. In fact, many people with psoriasis have colonization of fungus in their digestive system and on their skin. In a recent study, 21 out of 34 people with psoriasis were found to have *Candida albicans* in the spaces between their fingers or toes, and the majority were also affected by fungi from the Tinea family.[67] Another study looked at stool samples of people with psoriasis and other skin disorders. Researchers found a high number of disease-producing microbes, predominantly yeasts, in the colon. Treatment for yeast infection corresponded with a decrease in skin inflammation.[68]

Elimination diets and hypersensitivity testing have also produced profound results. People with psoriasis have high levels of IgE antibodies, which indicate an allergic component. Intestinal dysbiosis predisposes people to food and environmental sensitivities, so people with hypersensitivities need to heal the intestinal lining by taking appropriate bacterial supplements.

Poor liver function may contribute as well. Liver function profile tests and the metabolic screening questionnaire can help you determine liver function, and the metabolic screening questionnaire can also be used to follow your progress. Incorporate a detoxification program with an elimination/provocation diet to determine which foods may trigger your psoriasis (see Chapter 9).

Alcohol consumption contributes to psoriasis, since alcohol contains many toxic substances, which stress an overburdened liver. *Candida albicans* (yeast) thrive when beer and wine are consumed. Even one glass can provoke symptoms. Alcohol also increases intestinal permeability.

The causes and treatment of psoriasis are complex. Successful treatment must encompass several approaches reflecting its complexity. Look for underlying causes and develop a personal program based on your needs.

Functional Laboratory Testing

1. Food and environmental sensitivity testing, IgE and IgG4 testing.
2. Candida testing (either blood or stool).
3. Liver function profile.
4. Intestinal permeability testing.
5. Blood testing for vitamin and mineral status.

Healing Options

Elimination/provocation diet: Explore the relationship between your psoriasis and food and environmental sensitivities through laboratory testing and the elimination/provocation diet (see discussion in Chapter 7). For best results work with a nutritionist or physician who is familiar with food sensitivity protocols.

Multivitamin with minerals: Take a good quality multivitamin with minerals every day. Look for a supplement that contains at least 25,000 IU vitamin A, 400 IU vitamin D, 400 IU vitamin E, 800 mcg folic acid, 200 mcg selenium, 200 mcg chromium and 25-50 mg zinc. Each of these nutrients has been shown to be deficient in people with psoriasis.

Stress management: Flareups of psoriasis often occur after a stressful event. Because stress has to do with our own internalization of an event, even a mildly stressful situation can trigger psoriasis. Learning stress modification techniques can change your attitudes about stressful situations, allowing you to let them roll by more easily. In a recent study four out of 11 people showed significant improvement in psoriatic symptoms with meditation and guided imagery.[69] Hypnotherapy, biofeedback and walks in nature are other effective tools. Regular aerobic exercise is a powerful stress reducer.

Fish oils/EPA/DHA: Eating fish or taking fish oils has been shown to have an anti-inflammatory effect on psoriasis. Fatty acids contribute to healthy skin, hair and nails, and fish oils promote production of anti-inflammatory prostaglandins.[70] It is also possible that fish oils potentiate the activity of vitamin D and sunlight. Dosage: Eat cold water fish—salmon, halibut, mackerel, sardines, tuna and herring—two to four times per week or take EPA/DHA capsules under a physician's supervision. Fish oils increase blood clotting time so use them carefully.

Sunlight and vitamin D: Sunlight stimulates our bodies to manufacture vitamin D, which has been shown to be an effective treatment for psoriasis.[71] In general, slow tanning improves psoriasis, with sunshine and sunlamps prescribed as part of standard therapy. A recent study, done in Israel at the Dead Sea, long renowned for its treatment of psoriasis, showed that natural sunlight stimulated significant improvement in disease activity. One group was just given sunlight therapy, and the other received additional therapy in mud packs and sulfur baths. Both groups showed significant improvement in skin symptoms and with psoriatic arthritis, where present.[72]

Milk thistle/silymarin: Extracts of the herb milk thistle have been used since the fifteenth century for ailments of the liver and gallbladder. Milk thistle, also known as silymarin, contains anti-inflammatory flavonoid complexes that promote the flow of bile and help tone the spleen, gallbladder and liver. An excellent liver detoxifier, milk thistle has also been shown to have a positive effect on psoriasis. Dosage: 3 to 6 capsules of 175 mg standardized 80 percent milk thistle extract daily with water before meals.

Zinc: Zinc is necessary for maintenance and repair of skin, immune function and healing. Copper and zinc compete for the same receptor sites during absorption. When zinc is deficient, copper is usually elevated. This is true for people with psoriasis.

Many studies have determined that people with psoriasis have lower levels of zinc than people in control groups.[73] However, studies using oral zinc supplementation haven't always shown a clear improvement in psoriasis, though such studies have been of short duration—only six to ten weeks. Even though they didn't show improvement in the skin, they did show improvement in immune function and dramatic improvement in joint symptoms. It's possible that either zinc needs to be used along with other nutrients, or the timeframe of these studies was too brief to see improvement. Dosage: 50 mg zinc picolinate (an easily absorbed form of zinc) daily.

Selenium: Many studies have shown that people with psoriasis are deficient in selenium.[74] Selenium is part of a molecule called glutathione peroxidase which protects against oxidative damage (free radicals). Giving supplemental selenium to people with psoriasis showed an increase in glutathione peroxidase levels and improvement in immune function, though not an improvement in skin condition.[75] However, they were studies of short duration with selenium the only supplement. This underscores the concepts of patience when using natural therapies and of using more than one nutrient or approach at a time. Dosage: 200 mcg daily, which you can get in a good multivitamin. Selenium can be toxic, so more is not necessarily better.

Saccharomyces boulardii: *Saccharomyces boulardii* is a cousin to baker's yeast. It has been shown to raise levels of secretory IgA, which are low in psoriatic arthritis and psoriasis. Dosage: 1 acidophillus/bifidus supplement containing *Saccharomyces boulardii*.

Topical creams: Many topical creams, oils and ointments help psoriasis. Capsaicin, a cayenne pepper cream, helped 66 to 70 percent of the people who used it in a recent trial. The main side effect was that of a burning feeling associated with chili peppers, which quickly subsided.[76] Vitamins A and E have also been used topically with success; one physician alternates them, one each day. Creams containing zinc are also effective, as are salves containing sarsaparilla. Golden seal ointment or oral supplements can also be helpful.

Honduran sarsaparilla: Sarsaparilla, a flavoring in root beers and confections, has proven to be effective in psoriasis, especially the more chronic, large-plaque forming type. Sarsaparilla binds bacterial endotoxins.[77] Dosage: 2-4 tsp liquid extract (1:1) daily; 250-500 mg solid extract (4:1) daily.

Lecithin/phosphatidylcholine: Lecithin was used in a 10-year study from 1940 to 1950. People consumed 4 to 8 tablespoons of lecithin daily, along with small amounts of vitamins A, B1, B2, B5, B6, D, thyroid and liver preparations, and creams. Out of 155 patients, 118 people responded positively.

Lecithin-rich foods include soybeans, wheat germ, nuts, seeds, whole grains, eggs and oils from soy, nuts and seeds. Lecithin granules can be purchased in health food stores and added to foods as a cooking ingredient. Lecithin can also be purchased in capsule form, as can the active ingredient in lecithin, phosphatidyl choline. Dosage: 4-8 tbsp daily or 1-4 capsules of phosphatidyl choline.

SCLERODERMA

Scleroderma is a connective tissue disease characterized by a thickening and loss of elasticity in the skin, joints, digestive tract (especially in the esophagus), lungs, thyroid, heart and kidney. Mild or severe, it can flare up and subside in intensity. There are two forms of the disease: localized to one or two locations or generalized throughout the body. The most common initial complaint is loss of circulation in toes or fingers (Raynaud's syndrome), characterized by swelling and a thickening of skin.

Joint pain is an early symptom. As the disease progresses, the skin becomes taut and shiny, with the face becoming masklike. There may be red blotches on the skin where capillaries have broken. Small calcifications occur under the skin on the fingers. People become malnourished and may need supplemental foods or total parenteral feeding (tube feeding).

Scleroderma has definite digestive components. The esophagus is usually inflamed. The esophageal sphincter becomes stiff and loses elasticity which causes gastric juices to go up into the esophagus, burning the lining. People who have been on antacid medications are likely to have a yeast infection in their esophagus. Malabsorption leads to poor movement, dysbiosis and semi-obstructions in the small intestines. Small bowel bacterial overgrowth is common due to a loss of peristaltic function in the intestines. Use of steroid medications increases the likelihood of yeast infections in the digestive tract. Treatment with fluconazole will cure the infection temporarily but doesn't change the fact there is a loss of movement in the area.[78] Small bowel overgrowth must be routinely monitored and treated if an infection is present.

There is no single known cause of scleroderma. There is evidence that prolonged exposure to silica, silicone and chemical solvents significantly increases the risk of developing scleroderma. (Another possible association is in workers with repetitive hand and arm vibration.[79]) In a small study, 44 women and 6 men went through extensive testing and examination to see if there was a relationship between their work and autoimmune disease. They had been working in a factory for 6 years on average that produced scouring powder with a high silica content. Thirty-two, or 64 percent, showed symptoms of a systemic illness, six with Sjögren's syndrome, five with scleroderma, three with systemic lupus, five with a combination syndrome, and 13 who didn't fit into any definite pattern of disease. Seventy-two percent had elevated ANA, an indicator of autoimmune connective tissue diseases. The conclusion was that workers who are continually exposed to silica have a high probability of developing an autoimmune problem.[80] Silicone breast implants also significantly raise the risk of developing scleroderma.[81]

Natural therapies can work along with medical therapies for scleroderma. Infections must be treated, and beneficial flora given. Nutrients that help with collagen maintenance and repair are essential to help prevent loss of elasticity in skin and organs. Foods and supplements which help reduce production of arachidonic acid will reduce inflammation and pain. Good quality oils, fish, nuts and seeds work in this way. It's also important to increase circulation and oxygen supply to the tissues. Finally, a nutrient-dense food plan must be developed that works to offset the problems of malnutrition which are common.

Functional Laboratory Testing

1. Methane breath test for small bowel bacterial overgrowth.
2. Elisa IgE and IgG testing for food and environmental sensitivities.
3. Candida testing.
4. Comprehensive digestive and stool analysis.
5. Liver function profile.
6. Testing for silicone sensitivity should be done on an annual basis for women with breast implants. Saline implants have a silicone casing which may also cause problems.

Healing Options

Treat infections: Small bowel infections, esophageal candida and other infections are likely to reoccur. No specific research has been done to show that use of supplementation with flora and other natural therapies can help with reoccurrence, but they do help to boost the immune system. You may be able to keep the infection at bay with use of colloidal silver, grapefruit seed extract or garlic capsules. Each of these substances has wide antimicrobial properties, low toxicity and a low incidence of negative side effects. Your physician will probably use antibiotics.

Acidophilus and bifidobacteria: These help rebalance normal digestive flora. Dosage: one to two capsules or 1/4-1/2 tsp two or three times daily.

Saccharomyces boulardii: *Saccharomyces boulardii* is a cousin to baker's yeast, which has been shown to raise levels of secretory IgA. Saccharomyces has also been shown to be effective in combating small bowel infections caused by *Clostridium difficile*, the main bacteria to be found in small bowel overgrowth. If your methane breath test shows a small bowel infection, you'll want to take supplements of *S. boulardii*. Dosage: 1 to 2 capsules three times daily between meals.

Elimination/provocation diet: Be certain to explore the relationship between your scleroderma and food and environmental sensitivities through laboratory testing and the elimination/provocation diet. Follow directions in Chapter 7. For best results work with a nutritionist or physician who is familiar with food sensitivity protocols.

Diet: People with scleroderma are usually malnourished. So eat at least five servings daily of fruits and vegetables and as many organic and natural foods as possible. You may want to supplement your diet with nutrient-rich protein powder drinks available at health food stores. To eliminate all foods that don't contribute to your nutritional well-being, reread Chapter 8 and use the food chart. Make changes gradually.

Multivitamin with minerals: Poor diet, loss of movement in the digestive tract, loss of elasticity of the organs, infections and medications all contribute to the malabsorption of nutrients. Selenium and vitamin C deficiencies are common in people with scleroderma. At least 17 nutrients are essential for formation of bone and cartilage, so it's important to find a supplement that supports these needs. Look for a supplement that contains 10,000 IU vitamin A, 800-1000 mg calcium, 400-500 mg magnesium, 400 IU vitamin E, at least 250 mg vitamin C, 50 mg vitamin B6, 15-50 mg zinc, 5-10 mg manganese, 1-2 mg copper and 200 mcg selenium in addition to other nutrients. Follow the dosage on the bottle to get nutrients in appropriate amounts.

Vitamin C: Vitamin C is vital for formation of cartilage and collagen, which is a fibrous protein that forms strong connective tissue necessary for bone strength. Vitamin C also plays an important role in immune response, helping protect us from disease-producing microbes. Vitamin C also inhibits formation of inflammatory prostaglandins, helping to reduce pain, inflammation and swelling. If you have candida or bacterial overgrowth, vitamin C can boost your body's ability to defend itself. Vitamin C is also an antioxidant, needed to counter free-radical formation noted in sclerotic conditions. Dosage: 1-3 grams daily in an ascorbate or ester form.

Glycosamine sulfate: Glycosamine sulfate is used therapeutically to help repair cartilage, reduce swelling and inflammation and restore joint function. Green-lipped mussels are a rich source of glycosaminoglycans. Use of glycosamine sulfate has no associated side effects. Dosage: 500 mg two to four times daily.

Licorice: DGL (deglycyrrhized) licorice helps heal mucous membranes by increasing healing prostaglandins that promote mucus secretion and cell proliferation. Licorice also enhances the blood flow and health of intestinal tract cells. Be sure to use DGL licorice to avoid side effects caused by whole licorice. Dosage: Chew 2 tablets three to four times daily.

Slippery elm bark: Slippery elm bark has demulcent properties so it's gentle and soothing to mucous membranes. It has been a folk remedy for both heartburn and ulcers in European and Native American cultures and was used as a food by Native Americans. Slippery elm bark can be used in large amounts without harm. Drink as a tea or chew on the bark. Recipe for tea: Simmer 1 tsp of slippery elm bark in two cups of water for 20 minutes and strain. Sweeten if you wish, and drink freely. You can also purchase slippery elm lozenges at health food stores and some drug stores.

Glutamine: Although I was unable to find any references for use of glutamine to heal the esophagus, it makes theoretical sense. The digestive tract uses glutamine as a fuel source and for healing. It is effective for healing stomach ulcers, irritable bowel syndrome and ulcerative bowel diseases, and it is likely to be useful in the upper GI tract as well. Dosage: Begin a one-month trial with 8 grams daily in divided doses. If it's helpful, continue.

Ginger: Ginger can provide temporary relief, it has some anti-inflammatory properties, and it can help expel gas. Ginger can be used as an ingredient in food or taken as a supplement. Make a tea with 1/2 tsp of powdered ginger or a 1 1/2 tsp of fresh ginger per cup of boiled water. Steep for ten minutes and drink. If you'd like, sweeten it with honey. Cook with ginger and use it freely.

Meadowsweet herb: Also a demulcent, meadowsweet soothes inflamed mucous membranes. To use as a tea, take 1-2 tsp of the dried herb in one cup of boiled water. Steep for 10 minutes, and sweeten with honey if you like. Drink 3 cups daily.

Gamma linolenic acid: One gram of evening primrose oil was given to four women with scleroderma three times daily for one year. They experienced a reduction in pain, improved skin texture and healing of sores; red patches on skin due to broken capillaries were much improved. The researchers suggest that 6 grams daily may be of greater benefit.[82] Dosage: 3-6 g of evening primrose oil, borage oil or flaxseed oil daily.

Omega 3 fatty acids/fish oils: Fish oil capsules reduce morning stiffness and joint tenderness. Similar results can be obtained by eating fish high in EPA/DHA—salmon, mackerel, halibut, tuna, sardines, and herring—two to four times a week. (For more on Omega 3 fatty acids/fish oils, see discussion under Osteoarthritis.) Fish oils increase blood clotting time and should not be used by people with hemophilia or those who take anti-coagulant medicines or aspirin regularly. Dosage: It's easiest for most people to eat fish two to four times each week. High dosages in capsule form should be monitored by a physician.

Bromelain: Bromelain is an enzyme from pineapple that acts as an anti-inflammatory in much the same way that evening primrose, fish oils and borage oils do. It interferes with production of arachidonic acid, which reduces inflammation. It also prevents platelet aggregation and interferes with growth of malignant cells. Bromelain can be taken with meals as a digestive aid, but as an anti-inflammatory it must be taken between meals. Dosage: 500-1000 mg two to three times daily between meals.

Quercetin: Quercetin is the most effective bioflavonoid in its anti-inflammatory effects. It can be used to reduce pain and inflammatory responses and for control of allergies. Dosage: 500-1000 mg two to four times daily.

Co-Enzyme Q10: CoQ10 is an enzyme that works to enhance cell function. It is necessary for energy production, immune function and repair and maintenance of tissues. Widely used in Japan for heart disease, it has been researched as an antitumor substance. Dosage: 60-100 mg daily. Trial period: two months.

Part 6

Resources

Directory

Laboratories

The tests described earlier in this book are all performed by the laboratories listed below, which also offer a wide variety of other useful functional tests. The tests listed here are not the ordinary tests you will find in your local medical lab or hospital, although many tests that were originally only performed by these laboratories have found their way into local and national labs. Unless your physician is nutritionally oriented, he or she may be unfamiliar with many of these tests. Don't let that deter you from asking your physician to order any tests you feel may be appropriate for you and your condition. Your physician can call these labs for complete information packages and test kits. These labs also provide competent staff to assist your doctor with interpretation of the results.

Diagnos-Techs, Inc.
Clinical & Research Laboratory
6620 S. 192nd Place, J-104
Kent, WA 98032
800-878-3787
206-251-0596

Provides many laboratory tests, including parasitology testing, Candida, Helicobacter, Intestinal Permeability, Digestion Efficiency, Secretory IgA, Gastric pH, Adrenal Stress Profile, DHEA levels, Free Radical/Oxidative Stress Markers, Female Hormone Panels and liver function testing.

Doctor's Data Inc. & Reference Laboratory
P.O. Box 111
30W101 Roosevelt Rd.
West Chicago, IL 60185
708-231-3649
800-323-2784
Fax: 708-231-9190

Provides many laboratory tests, including Hair Mineral testing, Blood Mineral Analysis, Urine and Blood Complete Amino Acid testing, Mercapturic Acid testing, D-Glucaric Acid, Functional Folic Acid Result, Methylmalonic Acid(B12 status), Diet Analysis.

Great Smokies Diagnostic Laboratory
18A Regent Park Blvd.
Asheville, NC 28806
704-253-0621
800-522-4762

Provides CDSA, Lactose Breath testing, Parasite testing, Liver detoxification pathway testing, Intestinal Permeability testing, Small Bowel Inflammation testing, Helicobacter pylori, Secretory IgA by saliva.

Immuno Laboratories Inc.
1620 W. Oakland Park Blvd.
Ft. Lauderdale, FL 33311
305-486-4500
800-231-9197
Fax: 305-739-6563

Provides IgE Airborne & Food Allergy Assay, IgG Food Sensitivity Assay, Candida albicans assay, Epstein Barr Virus profile, Helicobacter Pylori assay, and Essential Metabolics Analysis through SpectraCell Labs.

Immunosciences Lab., Inc.
1801 La Cienega Blvd.
Los Angeles, CA 90035
310-287-1884

Provides tests designed to detect immunological injury due to environmental factors. They test for Comprehensive immune function, Lymphocyte immune testing, Natural Killer Cell activity, Immunotoxicology, Silicone related immune panels,

Chronic fatigue panels, Gastrointestinal evaluation, viral antibodies, parasite antibodies, Candida antibodies, bacterial antibodies, Secretory IgA.

Meridian Valley Clinical Laboratory
24030 132nd Ave SE
Kent, WA 98042
206-639-0941
800-234-6825

Provides a wide variety of tests including DHEA screening, blood lectin serotypes, Elisa Allergy Tests, Blood Mineral Analysis, Urine Mineral Analysis, and Hair Mineral Analysis, CDSA, Essential Amino Acid testing, Adrenal Steroids, Parasitology, Essential Fatty Acids, Fractionated Estrogens.

Meta Metrix Medical Laboratory
5000 Peachtree Industrial Blvd., Suite 110
Norcross, GA 30071
800-221-4640

Provides a wide variety of tests, including Amino Acid Analysis, Fatty Acid Analysis, lipid peroxides, antioxidant vitamin status, Minerals/Blood/Urine/Hair, whole blood reduced carnitine, glutathione, & ATP, IgG and IgE food testing, IgE inhalant allergy testing, functional liver detoxification, Intestinal Permeability testing, plasma homocysteine, cysteine, total glutathione.

Monroe Medical Research Laboratory Inc.
P.O. Box 1, Rt. 17
Southfields, NY 10975
914-351-5134
Fax: 914-351-4295

Provides Amino Acid Fractionated/plasma & urine, Functional testing for Long term Vitamin Status, Enzymes, and Cofactors, Organic Acids, Fatty Acids, Catecholamines & Indoleamines, Min-

eral & Metal Analysis/RBC/Lymphocytes/Leukocytes/or urine, Anti-oxidant status, Mauve factor, Glucaric Acid, Superoxide dismutase, Vit. E/platelet, Catalase, Histamine, PABA(pancreatic function test), Mercury(DMPS challenge test).

Pacific Toxicology Laboratories
1545 Pontius Ave.
Los Angeles, CA 90025
310-479-4911
800-32TOXIC
Fax: 310-479-2894

Provides testing for environmental pollutants and contaminants which includes panels for: Solvents & Metabolites, Pesticides & Herbicides, Polychlorinated and Polybrominated Biphenyls(PCB's & PBB's), Heavy Metals, Metals, OSHA Compliance Panels.

Serammune Physicians Laboratory
1890 Preston White Drive, 2nd floor
Reston, VA 22091
703-758-0610
800-553-5472
Fax: 703-758-0615

Provides comprehensive food and environmental blood testing through the Elisa/Act Test. The Elisa/Act test measures over 300 foods, environmental chemicals, preservatives, mercury sensitivity and three major classes of yeasts. Comprehensive physician and patient guides provide life-style support for the lab results.

SpectraCell Laboratories, Inc.
515-Post Oak Blvd., Suite 830
Houston, TX 77027
713-621-3101
800-227-5227

Provides a test called the Essential Metabolics Analysis. It is a blood test that checks for functional status of 19 vitamins, minerals and metabolites to measure long-term nutritional status.

Nutritional & Herbal Products

Listed below are companies that sell various health related products. This list is short and cannot include the many fabulous companies that exist. This does not constitute an endorsement of any one company over another. Some companies sell to both professionals and stores, although not all products are available in stores. Other companies sell only to health professionals. Professional products are indicated with an *.

Advanced Medical Nutrition (AMNI)
2247 National Ave.
Hayward, CA 94545
510-783-6969
800-437-8888
Nutritional products

Biotics Research Corp.
P.O. Box 36888
Houston, TX 77236
800-231-5777
In Texas: 800-833-5699
Nutritional products

DNA Pacifica*
730 Summersong Lane
Encinitas, CA 92024
619-632-5382
800-632-5382
Nutritional products

Apothécure Pharmacy
13720 Midway Rd.
Suite 109
Dallas, TX 75244
214-960-6601
800-969-6601
Custom pharmaceutical and nutritional products.

Cambridge Nutraceuticals
1 Liberty Square, 10th fl.
Boston, MA 02109
617-695-9553
800-265-2202
Glutamine products

Ecclectic Institute
14385 SE Lusted Rd.
Sandy OR 97055
503-668-4120
800-332-4372
Herbal products

Hobon*
P.O. Box 8243
Naples, FL 33941
941-643-4636
800-521-7722

Karuna*
42 Digital Dr. Suite 7
Novato, CA 94949
415-3828-0147
800-826-7225

Metagenics
971 Calle Negocio
San Clemente, CA 92673
714-366-0818
800-692-9400
Nutritional, herbal, homeopathic, Chinese

Neo-Life
Image Awareness Corp.
1271 High St.
Auburn CA 95603
916-823-7092

Perque/Seraphim, Inc.*
1890 Preston White Dr.
Reston, VA 22091
703-758-0689
800-525-7372
Nutritional products

Homeopathic Educational Services
2124 Kittredge Ave.
Berkeley, CA 94704
415-649-0294
800-359-9051

The Key Company
1313 W. Essex
P.O. Box 3707
St. Louis MO 63122
314-965-6699
800-325-9592 USA
800-392-9039 Missouri
Nutritional products

Natren Inc.
3105 Willow Lane
Westlake Village, CA 91361
805-371-4737
800-992-3323

NutriCology, Inc.
Allergy Research Group
P.O. Box 489
San Leandro, CA 94577
510-639-4572
800-545-9960
Nutritional & herbal products

Prima
1010 Crenshaw Blvd.
Suite 170
Torrance, CA 90501
310-320-1132
Nutritional & herbal products

Professional Botanicals*
P.O. Box 9822
Ogden, UT 94409
801-621-3450
800-824-8181
Herbal products

Progressive Laboratories, Inc.
1701 Walnut Hill Lane
Irving, TX 75038
214-518-9660
800-527-9512

Systemic Formulas*
P.O. Box 1516
Ogden, UT
801-621-8840
Nutritional & herbal products

Tyler Encapsulations*
2204-8 N.W. Birdsdale
Gresham, OR 97030
503-661-5401
800-869-9705
Nutritional & herbal products

Professional & Technical Services
5200 SW Macadam Ave.
Suite 420
Portland, OR 97201
503-226-1010
800-648-8211
Nutritional & herbal products

Spectrum Naturals
133 Copeland St.
Petaluma CA 94952
707-778-8900
Essential fatty acids
Cooking oils

Thorne Research, Inc.*
P.O. Box 3200
Sandpoint, ID 83864
208-263-1337
800-228-1966

Werum Enterprises, Inc.*
P.O. Box 903
Shingle Springs, CA 95682
916-677-1988
800-822-6193
Nutritional & herbal products, books

Organizations

Nutritionally Oriented Physicians

American Academy of
Environmental Medicine
P.O. Box 16106
Denver, Co 80216

American Chiropractic
Assoc.
1701 Clarendon Blvd.
Arlington, VA 22209

American Assoc. of Naturopathic Physicians
2355 Eastlake Ave, Suite
322
Seattle, WA 98102
206-323-7610

American College of
Advancement in
Medicine
P.O. Box 3427
Laguna Hills, CA
USA: 800-532-3688
Calif.: 714-583-7666

American Holistic Medical
Assoc.
2727 Fairview Ave. E.
Seattle, WA 89102

Human Ecology Action
League
HEAL
P.O. Box 66637
Chicago, IL 60666
312-665-6575

Nutritionists/Nutritional Information

International & American
Associations of Clinical
Nutritionists
5200 Keller Springs Rd
Suite 410
Dallas, TX 75248

Nutrition for Optimal
Health Association
(NOHA)
P.O. Box 380
Winnetka, IL 60093

Acupuncture

American Assoc. of Acupuncture and Oriental Medicine
4101 Lake Boone Trail, Suite 201
Raleigh, NC 27607
919-787-5181

Biofeedback

Assoc. for Applied Psychophysiology and Biofeedback
10200 W. 44th Ave., #304
Wheat Ridge, CO 80033
303-422-8436

Herbalists & Herbs

American Botanical Council
P.O. Box 210660
Austin, TX 78720
512-331-8868

American Herbalists Guild
P.O. Box 1683
Sequel, CA 95073

Associations for Specific Illnesses

American Celiac Society
45 Gifford Ave
Jersey City, NJ 07304

Gluten Intolerance Group
26604 Dover Court
Kent, WA 98031

Crohn's & Colitis Foundation of America
3386 Park Ave. South
NY, NY 10016-7374
800-932-2423
212-685-3440

Celiac Disease
Celiac Sprue Assoc
2313 Rocklyn Drive #1
Des Moines, IA 50322

National Celiac Sprue Society
10 Larkspur Way #4
Natick, MA 01760

Reach Out for Youth with Ileitis and Colitis, Inc.
15 Chemung Place
Jerico NY 11753
516-822-8010

Pediatric Crohn's & Colitis Association, Inc.
P.O. Box 188
Newton, MA 02168
617-244-6678

Chronic Fatigue Syndrome Association
P.O. Box 220398
Charlotte NC 28222-0398
704-362-2343

Rheumatoid Disease Foundation
5106 Old Harding Rd.
Franklin, TN 37064
615-646-1030

American Liver Foundation
1425 Pompton Ave.
Cedar Grove, NJ 07009
800-223-0179
201-256-2550

Digestive Disease National Coalition
711 2nd St., NE, Suite 200
Washington, DC 20002
202-544-7497

Intestinal Disease Foundation, Inc.
1323 Forbes Ave, Ste. 200
Pittsburgh, PA 15219
412-261-5888

National Headache Foundation
5252 N. Western Ave.
Chicago, IL 60625
800-843-2256

CFIDS Buyer's Club
1187 Coast Village Rd. #1-280
Santa Barbara, CA 93108

Candida Research Information Foundation
P.O. Box 2719
Castro Valley, CA 94546
415-582-2179

Center for Digestive Disorders
25 N. Winfield Rd.
Winfield, IL 60190-1295
708-682-1600, Ext. 6493

International Foundation for Bowel Dysfunction
P.O. Box 17864
Milwaukee, WI 53217
414-964-1799

National Digestive Diseases Information Clearinghouse
Box NDDIC
9000 Rockville Pike
Bethesda, MD 20892
301-654-3810

Appendix

Health Hazard Appraisal

Your physician or nutritionist can obtain a copy of the complete appraisal from Metagenics and Tyler Encapsulations listed above under Nutritional & Herbal Products.

Suggested Reading List

Bacteria/Antibiotics/Parasites

Garrett, Laurie, *The Coming Plague*, Farrar, Straus & Giroux, New York, 1994

Louise Gittleman *Guess What Came to Dinner? Parasites & Your Health*, Avery Publ., Garden City Park, New York, 1993

Schmidt, Michael, DC, Sehnert, Keith MD, Smith, Lendon MD, *Beyond Antibiotics*, North Atlantic Books, Berkeley, CA 94705, 1993

Trenev, Natasha, Chaitow, Leon, *Probiotics*, Thorsens, 77-85 Fulham Palace Rd., Hammersmith, London W68JB, 1990

Candida/Yeast Infections

Crook, William MD, *The Yeast Connection*, 1985, and *The Yeast Connection and the Woman*, 1995, and Professional Books, P.O. Box 3494, Jackson, TN 38301

Trowbridge, John MD and Walker, Morton Ph.D., *The Yeast Syndrome*, Bantam Books, New York, 1986

Food & Environmental Allergies

Coca, Arthur, M.D., *The Pulse Test*, Arco Publ. Co., New York, 1978

Dadd, Debra Lynn, *The Nontoxic Home*, Jeremy Tarcher, Los Angeles, 1986

Golos, Natalie & Golos-Golbitz, Frances, *If This is Tuesday, It Must Be Chicken*, Keats Publ., New Canaan, CT, 1983

Randolph, Theron, MD, Moss, Ralph, Ph.D., *An Alternative Approach to Allergies*, Harper & Row, New York, 1989

Food & Nutrition

Ballantine, Rudolph MD, *Diet and Nutrition*, Himilayan Institute, Honesdale, PA, 1978

Colbin, Ann Marie, *Food & Healing*, Ballantine Books, New York, 1986

Erasmus, Udo, Ph.D., *Fats that Heal, Fats that Kill*, Alive Books, 7436 Fraser Park Dr., Burnaby BC, Canada, V5J5B9. 1993

Gottschall, Elaine, BA, M.Sc., *Food & the Gut Reaction*, The Kirkton Press, Kirkton, Ontario, Canada, 1986

Lappe, Frances Moore, *Diet for a Small Planet*, 1971 & 1991 and *Food First*, 1978, Ballantine Books, New York

May All Be Fed, William Morrow & Co., 1992

Price, Weston, *Diet and Physical Degeneration*, Keats Publ., New Canaan, CT, first publ. 1938

Shabert, Judy, MD, RD, Nancy Erlich, *The Ultimate Nutrient, Glutamine*, Avery Publ., Garden City Park, NY, 1994

Robbins, John, *Diet for a New America*, Stillpoint Publ., Walpole, NH 03609, 1987

Holistic Self-Care

Gardner, Joy, *The New Healing Yourself*, The Crossing Press, Freedom CA 95019, 1989

Murray, Michael, and Joseph Pizzorno, N.D.s, *The Encyclopedia of Natural Medicine*, Prima Publishing, P.O. Box 1260MP, Rocklin, CA 95677, 1991

Wellness

Ardell, Don, *14 Days to a Wellness Lifestyle*, Whatever Publishing, Inc. 158 E. Boithedale, Mill Valley, CA 94941, 1982

———, *High Level Wellness*, 10 Speed Press, P.O. Box 7123, Berkeley, CA 94707

Borysenko, Joan, Ph.D., *Minding the Body, Mending the Mind*, Bantam Books, New York, 1987

Ryan, Regina, and Travis, John, *The Wellness Workbook*, 10 Speed Press, P.O. Box 7123, Berkeley, CA 94707, 1981

REFERENCES

Chapter 1. Who Can Benefit from This Book?

1. *Digestive Diseases in the United States: Epidemiology and Impact*, U.S. Department of Health and Human Services, Public Health Service, NIH, Publication 94-1447, May 1994, p. 19.
2. Burkitt D, *Eat Right to Stay Healthy and Enjoy Life More*, Arco Publishing, New York, 1979, pp. 17-31; Murray M; Pizzorno J; *Encyclopedia of Natural Health*, Prima Publ., Rocklin, CA, p.576.
3. "Helicobacter pylori in Peptic Ulcer Disease," National Institutes of Health Consensus Statement, Feb 7-9, 1994, Volume 12, Number 1.
4. This questionnaire has been used with permission from Lyra Heller & Michael Katke, *Metagenics*, San Clemente, CA, 1985, and Corey Resnick, N.D. at Tyler Encapsulations, Gresham OR, 1995. I have adapted it to meet the specific needs of digestive wellness.

Chapter 2. The American Way of Life Is Hazardous to Your Health

1. Jaffe R; Donovan P; *Guided Health, A Constant Professional Reference*, USDA, 1988-9 data, Health Studies Collegium Publ., Reston, VA, pp. 3.12, 5.56. Sixth Special Report to the U.S. Congress on Alcohol and Health.
2. Kennedy SH; "Vitamin Supplements Win New Found Respect: Skeptics Are Changing Their Minds Over the Value of These Compounds," *Modern medicine*, July 1992;60:15-18; Hemilia H; "Vitamin C and Plasma Cholesterol, *Critical Reviews in Food Science and Nutrition* 1992;32(1):35-37; Jacques P; "Effect of Vitamin C on High-Density Lipoprotein Choles-

terol and Blood Pressure," *Journal of the American College of Nutrition*, 1992;11(2):139-144.
3. Pauling L; "Prevention and Treatment of Heart Disease, New Research Focus at the Linus Pauling Institute", *Linus Pauling Institute of Science & Medicine Newsletter*, March 1992;1.
4. Neher JO, Borkan JM; "A clinical approach to alternative medicine," *Archives of Family Medicine*, Oct. 1994;3:859-861.
5. Boris, J; Mandel FS; "Foods and additives are common causes of the attention deficit hyperactive disorder in children," *Ann Allergy*, 1994 May;72(5):462-8; Carter CM, et al.; "Effects of a new food diet in attention deficit disorder," *Arch Dis Child*, 1993 Nov;69(5):564-8.
6. "Exclusive Interview with Gary Gibbs, D.O., Authority on Food Radiation," *Health & Healing*, May 1995;2(5):5-6.
7. Wolfe SM; *Health Letter*, Washington, DC, The Public Citizen Health Research Group, 1989;5(7)1-5.

Chapter 3. A Voyage Through the Digestive System

1. Robbins J; *May All Be Fed*, William Morrow & Co., 1992, p. 26.
2. Tips, Jack, *The Liver Triad*, Apple A Day Publishing, 1989, p. 12.
3. Ibid.

Chapter 4. The Bugs in Your Body!: Intestinal Flora

1. Mitsuuoka T; "Intestinal Flora & Aging," *Nutrition Reviews* Dec. 1992; vol. 50(12): 438-46.
2. Galland L; "Dysbiosis & Disease", tape from Great Smokies Seminar 1993.
3. Sellars RL, "Acidophilus Products," *Therapeutic Properties of Fermented Milks*, Elsevier Applied Science, 1991, p. 102.
4. Chaitow L, Trenev N, *Probiotics*, Thorsons, 1990, pp. 28-31.
5. *Ibid.*, p, 57.
6. Sehnert K; *The Garden Within*, Health World Magazine, 1989, p. 9; Chaitow L., Trenev N., *op. cit.*, pp. 15, 57, 144.
7. Sellars, *op. cit.*, pp. 103-4.
8. Galland, Leo; "Solving the Digestive Puzzle", conference manual, Great Smokies Diagnostic Lab and HealthComm Interna-

tional, San Francisco, May 1995, p. 7; Kimmey MB; Elmer GM: et al; "Prevention of further recurrences of Clostridium difficile colitis with Saccharomyces boulardii", *Dig. Dis. Sci.* 1990 Jul;35(7):897-901; Plein K; Hotz J; "Therapeutic effects of Saccharomyces boulardii on mild residual symptoms in a stable phase of Crohn's disease with special respect to chronic diarrhea—a pilot study", *Gastroenterology* 1993 Feb; 31(2):129-34; Czerucka D; Roux I; Rampal P; "Saccharomyces boulardii inhibits secretagogue-mediated adenosine 3',5'-cyclic monophosphate induction in intestinal cells.", *Gastroenterology* 1994 Jan;106(1):65-72.

9. Sellars, *op. cit.*, pp. 103-04.
10. Carper, Jean, *Food Pharmacy*, Bantam Books, 1988, p. 124.
11. Telephone conversation with Natasha Trenev, Winter 1994
12. Telephone conversation with Corey Resnick, N.D., Winter 1994
13. Sehnert, *op. cit.*, p. 16.

Chapter 5. Dysbiosis: A Good Neighborhood Gone Bad

1. Bland J, *Preventive Medicine Update*, audiotape, May 1994.
2. For a lengthy but fascinating look at the world through the eyes of virologists, read *The Coming Plague*, by Lori Garrett, Farrar, Straus, Giroux, 1994.
3. Begley S; "The end of antibiotics," *Newsweek*, March 28, 1994, p 46-51.
4. McBride J; "Nutrient Deficiency Unleashes Jekyll-Hyde Virus," *Agricultural Research*, August 1994, vol. 42(8): 14-16. Review of research by M. Beck et al., University of North Carolina at Chapel Hill, N.C.
5. Tippett K; Goldman JD; "Diets More Healthful, But Still Fall Short of Dietary Guidelines," *Food Review, U.S. Department of Agriculture*, Jan-Apr 1994;17:(1):8-14.
6. *Ibid.*; "Providers Estimate One in Four Seniors Malnourished," *Nutrition Week*, April 30, 1993;3 (survey sponsored by the Nutrition Screening Initiative, American Dietetic Association, the National Council on the Aging and the American Academy of Family Physicians); Loosli AR, "Reversing Sports-Related Iron & Zinc Deficiencies", *The Physician & Sportsmedicine*, June 1993;21(6):70- 78.

7. Personal correspondence from Martin Lee of Great Smokies Laboratories, Spring 1995.
8. Mitsuoka, T; "Intestinal Flora & Aging," *Nutrition Reviews* Dec. 1992, vol. 50(12): 438-46.
9. Koelz HR, "Ulcer prevention during anti-rheumatism therapy and in intensive medicine", *Schweizs Rundsch Med Prax*, 1994 June 21;83(25-26):768-71; Bjarnason I; et al; "Side Effects of Non-steroidal anti-inflammatory Drugs on the Small & Large Intestine in Humans", *Gastroenterology*, 1993 June;104(6):1832-47.
10. Galland, *op. cit.*
11. Abraham Hoffer, preface to *The Yeast Syndrome* by John Trowbridge and Morton Walker, Bantam Books 1986.
12. Candida Questionnaires used with permission of William Crook, M.D., from his book: *The Yeast Connection and the Woman*, Professional Books, Jackson, TN, 1995.

Chapter 6. Leaky Gut Syndrome: The Systemic Consequences of Faulty Digestion

1. "The Leaky Gut," *Great Smokies Digest*, summer 1990, p. 4; "Gut Hyperpermeability", Serammune Physicians Lab, Volume 2, Number 1, January 1992; Trenev, Natasha, Lecture to California State Meeting IAACN, April 1994.
2. Galland, Leo, "Solving the Digestive Puzzle," Great Smokies Diagnostic Laboratory/HealthComm International Inc., Conference manual, San Francisco, May 1995, p. 10.
3. Ibid., page 11; Swanson, Mark; audiotape: "Comprehensive Digestive Stool Analysis & Intestinal Permeability", Great Smokies Diagnostic Lab., Spring 1993.
4. "Gut Hyperpermeability," Serammune Physicians Laboratory, Jan. 1992, vol. 2(1).
5. Marks, David and Laura, "Food Allergy: Manifestations, Evaluation and Management", Food Allergy, *Postgraduate Medicine*, Feb. 1, 1993:93(2):191-201.
6. *Ibid.*
7. *Guide to Health*, Immuno Laboratories, Inc., 1994; Jaffe, Russ, Serammune Physicians Lab, educational brochure.
8. *Elisa/Act Handbook*, Serammune Laboratories.

Chapter 7. Functional Medicine/Functional Testing

1. Gittleman, Louise, *Guess What Came to Dinner*, Avery Publishing, 1993.
2. Parasite Questionnaire used with permission from Louise Gittleman, *Guess What Came To Dinner?* Avery Publishing Group, Inc., Garden City Park, New York, 1993.
3. Lab manual, Great Smokies Diagnostic Laboratory.
4. Godiwala T et al., American College Gastroenterology, Oct.18, 1988, NYC.
5. Rider MS; Acterberg J; et al; "Effect of Immune System Imagery on Secretory IgA, *Biofeedback Self Regul* 1990 Dec;15(4); 317-33.
6. Application Guide, Functional Liver Testing, Great Smokies Diagnostic Laboratory, 1994.
7. "New Clinical Breakthroughs in the Management of Chronic Fatigue Syndrome, Intestinal Dysbiosis, Immune Dysregulation & Cellular Toxicity," by Jeffrey Bland, Health Comm, 1992.
8. "Bacterial Overgrowth of the Small Intestine," Application Guide, Great Smokies Laboratory, 1994.
9. Kirsch, Michael MD, "Bacterial Overgrowth," *The American Journal of Gastroenterology*, Vol 1990;85(3):231-37.
10. *Ibid.*

Chapter 8. Moving Toward a Wellness Lifestyle

1. Adapted from *Structured Exercises in Wellness*, Vol. 3, Nancy Loving Tubesing and Donald Tubesing, Whole Person Press, 1986. Idea originally adapted by Kent Beeler from *The Wellness Series I Booklet* by the Mennonite Mutual Aid of Goshen, Indiana, 1983.

Chapter 9. First Things First/Detoxification

1. *Ultrabalance Workbook*, 1988, Health Comm, Inc., p.111.
2. National Research Council, *Environmental Neurotoxicity*, National Academy Press, WA, DC 1992, as referenced in *New Clinical Breakthroughs*, by Jeffrey Bland, 1992-3.

3. For resource material on fasting see Paavo Airola's Book on *Juice Fasting* or Elson Haas, *Staying Healthy with Nutrition.*
4. This rice-based protein drink is available only through health professionals and is sold by Metagenics.
5. This questionnaire was used with permission of HealthComm International, Inc., 5800 Soundview Drive, Gig Harbor, WA 98335.
6. Stone, I, *The Healing Factor*, Grosset & Dunlap, New York, 1982, pp. 48, 152-162.
7. Sokol-Green N; *Poisoning Our Children*, Noble Press Inc. 1991.
8. Russell Jaffe, *Women and Children's Health Update*, 1994-5 Syllabus.

Chapter 10. Diet Means "Way of Life"

1. Erasmus, Udo, *Fats that Heal, Fats that Kill*, Alive Books, 1993, p. 109.
2. One of the best books I've read on this subject is: *Circle of Poison*, by D. Weir and M. Schapiro, Institute for Food & Development Policy, 2588 Mission St. San Francisco, CA 94110, 1981.
3. Smith, Bob, "Organic Foods vs. Supermarket Foods, Element Levels," Doctor's Data, West Chicago, IL., 1993.
4. Jaffe, R., Donovan, P., Guided Health, *op. cit.*, p. 5.46.
5. Schnohr, P., et al., "Egg Consumption and High-Density-Lippoprotein Cholesterol," *Journal of Internal Medicine*, 1994;235:249-251.
6. Shinitsky, M, "Egg Consumption, Serum Cholesterol & Membrane Fluidity," *Biomembranes and Nutrition*, 1989;195:391-400.
7. Schauss, Alex, "Dietary Fish Oil Consumption & Fish Oil Supplementation," *A Textbook for Natural Medicine*, Pizzorono and Murray editors, 1991, p. 1-7; Carper, J., *The Food Pharmacy*, Bantam Books, 1989, p 55.
8. Robbins, John, *May All Be Fed*, William Morrow Publ., 1992, p. 35.

Chapter 12. Natural Therapies for Common Digestive Problems

1. Siblerud RL; "Relationship Between Mercury from Dental

Amalgam and Oral Cavity Health," *Annals of Dentistry*, Winter 1990, NY Academy of Dentistry.

2. Shibasaki T; "The Relationship of Nutrition and Dietary Habits to Gingivitis, Dental Calculus Deposit, and Dental Plaque Adhesion in High School Students," *Shoni Shikagaku Zasshi* 1989;27(2):415-26.
3. Seigel MA; Balciunas BA; "Medication can induce severe ulcers," *J Am Dental Assoc.* 1991 Sep;122(10):75-77.
4. Nolan A; McIntosh WB; Allam BF; Lamey PJ; "Recurrent apthous ulceration: Vitamin B1, B2 and B6 status and response to replacement therapy," *J. Oral Path. Medicine* 1991 Sept; 20(8):389-91; Palopoli J; Waxman J; "Recurrent apthous stomatitis and Vitamin B12 deficiency," *Southern Medical Journal* 1990 April;83(4):475-7; Porter SR; Scully C; Flint S; "Hematologic status in recurrent apthous stomatitis compared with other oral disease," *Med Oral Pathol* 1988 July;66(1):41-4.
5. Malstrom, M, Salo OP; Fyhrquist F, "Immunogenetic markers and immune response in patients with recurrent oral ulceration," *Int. J. Oral Surgery*, 1983 Feb;12('1):23-30; Ofarrelly C; O'Mahony C;Graeme-Cook F; Feighery C; et al, "Gliadin antibodies identify gluten-sensitive oral ulceration in the absence of villous atrophy," *J. Oral Path. Medicine* 1991 Nov;20(10):476-8; Wray D, "Gluten-sensitive recurrent apthous stomatitis," *Digestive Disease Sciences*, 1981 Aug;26(8):737-40; Walker DM; Dolby AE; Mead J; et al., "Effect of gluten-free diet on recurrent apthous ulceration," *British Journal Dermatology* 1980 July;103(1):111.
6. Wang SW, Li HK; He JS; Yin TA "The trace element zinc and apthosis. The determination of plasma zinc and the treatment of apthosis with zinc." *Rev Stomatol Chir Maxillofac* 1986:87(5):339-43.
7. Challacombe SJ, "Hematological abnormalities in oral lichen planus, Candidiasis, leukoplakia and non-specific stomatitis," *Int J Oral Maxillofac Surg* 1986 Feb;15(1):72-80.
8. Drinka PJ; Langer E; Scott L; Morrow F; "Laboratory measurements of nutritional status as correlates of atrophic glossitis," *J Gen Internal Medicine* 1991 Mar-Apr;6(2):137-40.
9. Husebye E; Skar V; Hoverstad T; Melby K; "Fasting Hypochlorhydria with Gram Positive Gastric Flora is Highly Prevalent in Healthy Old People," *Gut*, 1992 Oct;33:1331-7; Wright

J; *Dr. Wrights Guide to Healing with Nutrition,* Rodale Books, 1984, p. 33.

10. R. Russell, *Clinical Pearls* 1991-2, p. 43.
11. National Institutes of Health Consensus Statement, "Helicobacter pylori in Peptic Ulcer Disease," Feb 1994, Volume 12, Number 1.
12. Wilhelmsen I; Berstad A; "Quality of Life and Relapse of Duodenal Ulcer before and after Eradication of Helicobacter pylori," *Scand. J. Gastroenterology* 1994 Oct;29(10): 874-9.
13. Telephone conversation with Leo Galland, MD, Winter 1995.
14. Kimikazu I; Kiyonana J; Ishikawa M; "Studies on Gamma-Oryzanol II: The Anti-Ulcerogenic Action, Research Institute, Otsuka Pharmaceutical Co., Ltd, Tokushima 771-01; Resnick C; "The Effects of Gamma-Oryzanol on Ulcers, Gastritis, Hyperlipidemias, and Menopausal Disorders," *Research Review,* Tyler Encapsulations, 1993.
15. Maruyama K; Kashiwzaki K; "Clinical Trial of Gamma-oryzanol on Gastrointestinal Symptoms at 375 Hospitals," Dept. of Internal Medicine, Keio University, Japan; Yoshinari, T; "Usefulness of Hi-z fine granule (gamma-oryzanol) for the treatment of Autonomic instability in Gastrointestinal System," *Shinyaku to Rinsho,* 225(3):56, 1976; Minakuchi C; et al;"Effectiveness of gamma-oryzanol on Various Gastrointestinal Complaints," *Shinyaku to Rinsho,* 25(10):29, 1976.
16. Sano-Gastril is marketed in this country by Nutri-Cology/Allergy Research Group. This is not an endorsement of a product, rather it is the only product of its type.
17. Bland J; *Glandular Based Food Supplements: Helping to Separate Fact from Fiction,* Bellevue-Redmond Medical lab. Inc., Bellevue, WA, p. 3
18. Barnard, N; "Fiber, Health, Nd Research; The Work of Dennis Burkitt, M.D.," *PCRM Update,* May-June, 1990;1-9.
19. "Fasting May Cause Stones," *Medical Tribune,* July 25, 1991;13; Capron JP, et al; "Meal Frequency and Duration of Overnight Fast: A role in Gall-stone formation?," *British Medical Journal,* 1981; 283:1435
20. Telephone conversation with Leo Galland, M.D., June 1995.
21. Moerman, C; "Dietary Risk Factors for Clinical Diagnosed Gallstones in Middle-Aged Men: A 25 Year Follow-up Study," *Annals of Epidemiology,* 1994;248-254; Moerman, C; "Dietary

Sugar Intake in the Etiology of Biliary Tract Cancer," *International Journal of Epidemiology*, 1993;22(2):207- 213.

22. Breneman J:, "Allergy Elimination Diet as the Most Effective Gallbladder Diet," *Annals of Allergy*, 1968 Feb;26(2):83-7.
23. Capper Wm., et al; "Gallstones, gastric secretion, and flatulent dyspepsia," *Lancet*, Feb 25, 1967;1:413-15.
24. *Merck Manual*, 16th Edition, Merck Research Laboratories, 1992, p 924; Tuzhilin SA, et al; "The Treatment of Patients with Gallstones by Lechitnin," *American Journal of Gastroenterology*, 1976. 65:231; Murray M, Pizzorno; *Encyclopedia of Natural Medicine*, Prima Publ., Rocklin CA, 1991, p. 235.
25. Simon JA; "Ascorbic Acid and Cholesterol Gallstones," *Med Hypothesis* 1993 Feb;40(2):81- 4; Chen L; "Bile acid pool in the formation of pigment stones: An Experimental Study," *Chung Hua Wai Ko Tsa Chih* 1992 Aug; 30(8):496-8.
26. Di Prisco, MC; "Possible Relationship Between Allergic Disease and Infection by Giardia Lamblia," *Annals of Allergy*, March 1993;70:210-212; Sky PR, "Of Parasites and Pollens," *Discover*, Sept. 93:56-62.
27. Phillipson JD and Wright CW; "Medicinal Plants in Tropical Medicine: Medicinal plants against protozoal diseases," Royal Society of Tropical Medicine and Hygiene, London June 221, 1990, *Transactions of the Royal Society of Tropical Medicine and Hygiene* 1991;85:18-21.
28. Lazzari R; et al; "Sideropenic anemia and celiac disease," *Pediatr Med Chir* 1994 Nov-Dec;16(6):549-50.
29. Catassi C; et al; "Coeliac diseases in the year 2000: Exploring the Iceburg," *Lancet*, 1994;343:200-3; Lazzari R; et al; "Sideropenic anemia and celiac disease," *Pediatr Med Chir* 1994 Nov-Dec;16(6):549-50.
30. "Constipation" fact sheet, National Institutes of Health, National Institute of Diabetes and Digestive & Kidney Diseases, Publication #92-2754, July 1992, p. 570.
31. Rogers, S; "Chemical Sensitivity: Breaking the Paralyzing Paradigm: How Knowledge of Chemical Sensitivity Enhances the Treatment of Chronic Disease," *Internal Medicine World Report*, 1992;7(8):13-41.
32. Whitehead WE; "Biofeedback Treatment of Gastrointestinal Disorders," *Biofeedback and Self-Regulation*, 1992;17(1):59-76; Bleijenberg G and Kuijpers HC; "Biofeedback Treatment of

Constipation: A Comparison of Two Methods," *American Journal of Gastroenterology*, July 1994;89(7):1021-1026.

33. Wilson JM; "Hand Washing Reduces Diarrhea Episodes: A study in Lombok Indonesia," *Transactions of the Royal Society of Tropical Medicine and Hygiene,* 1991;85:819-821.
34. Mendeloff AI; Everhart JE; "Diverticular Disease of the Colon," *Digestive Diseases in the United States: Epidemiology and Impact*, U.S. Dept. of Health & Human Services, National Institutes of Health, 94-1447, May 1994, p. 553; Ozick LA; Salazar C; Donelson SS; "Pathogenesis, Diagnosis and Treatment of Diverticular Disease of the Colon," *Gastroenterologist*, Dec. 1994, vol. 2 (4): 299-310.
35. Aldoori WH, et al; "A Prospective Study of Diet and the Risk of Symptomatic Diverticular Disease in Men," *American Journal of Clinical Nutrition*, 1994;60:757-64.
36. Baran E; Dupont C; "Modification of Intestinal Permeability During Food Provocation Procedures in Pediatric Irritable Bowel Syndrome," *Journal of Pediatric Gastroenterology & Nutrition*, 1990;11:72-77; Bell IR; et al; "Symptom & Personality Profiles of Young Adults from a College Student Population with Self-Reported Illness from Foods and Chemicals," *Journal of the Amer College of Nutrition,* 1993;12(6):693-702.
37. Vernia P; et al; "Lactose Intolerance and Irritable Bowel Syndrome: Relative Weight in Inducing Abdominal Symptoms in High Prevalence Area," *Gastroenterology*, April 1992;102(4):Part II/A530; Rumessen JJ; "Functional Bowel Disease: The Role of Fructose and Sorbitol," *Gastroenterology*, 1991;101:1452-1460; Born P; et al; "Fructose Malabsorption and Irritable Bowel Syndrome," *Gastroenterology*, 1991;101(5):1454; Fernandez-Banares F, et al; "Role of Fructose-Sorbitol Malabsorption and Irritable Bowel Syndrome," *Gastroenterology*, 1991 Nov;101(5)1453-4; Francis CW; Whorwell PJ; et al., "Bran and Irritable Bowel Syndrome: Time for Reappraisal," *The Lancet,* July 2, 1994;334:339-40.
38. Hoffman R; 7 *Weeks to a Settled Stomach*, Pocket Books, New York,1990, pp. 209, 211.
39. Francis CW; Whorwell PJ; et al., "Bran and Irritable Bowel Syndrome: Time for Reappraisal," *The Lancet,* July 2, 1994;334:339-40; Hotz J; Plein K; "Effectiveness of plantago seed husks in comparison with wheat bran on stool frequency

and manifestations of irritable colon syndrome with constipation," *Med Klin*,1994 Dec 15;89(12):645-51.

40. Tomas-Ridocci M; et al; "The Efficacy of Plantago ovata as a Regulator of Intestinal Transit," *Rev Esp Enferm Dig*,1992 Jul;82(1):17-22.
41. Gaby A; Wright J; "Ulcerative Colitis," *Nutrition & Healing*, Jan 1995, v.2, (1); Murray M; Pizzorno J; *The Encyclopedia of Natural Medicine*, Prima, Rocklin, CA, 1991, p. 241.
42. Gottschall E; *Food and the Gut Reaction*, Kirkton Press, Kirkton, Ontario, Canada, 1986; Fernandez-Banares, et al, "Enteral Nutrition as a Primary Therapy in Crohn's Disease," *Gut*, 1994; Suppl 1:s55-s59; Nellist CC; "Elemental Diet Therapy a good option for Crohn's," *Family Practice News*, March 1, 1994;7; Bjarnason I, "Intestinal Permeability," *Gut*, 1994; Suppl. 1:s18-s22
43. "Crohn's disease linked to measles," *Medical Tribune*, May 13, 1993;10.
44. Gaby A; Wright J; "Ulcerative Colitis," *Nutrition & Healing*, Jan 1995, v.2, (1).
45. Gryboski JD; "Ulcerative colitis in children 10 years old or younger," *Journal of Pediatric Gastroenterology Nutr*; 1993 Jul;17(1):24-31.
46. Seigel J; "Inflammatory bowel disease: Another possible effect of the allergic diathesis," *Annals of Allergy*, 1981 Aug;47(2):92-4.
47. Hanauer SB, Peppercorn MA; Present DH; "Current Concepts, New Therapies in IBD," *Patient Care*, August 1992.
48. Boyko EJ; et al; "Increased risk of inflammatory bowel disease associated with oral contraceptive use," *American Journal of Epidemiology*, 1994;140(3):268-278.
49. Gottschall E; *Food and the Gut Reaction*, Kirkton Press, Kirkton, Ontario, Canada, 1986, also published more recently as: *Breaking the Vicious Cycle.*
50. Aslan A; Triadafilopoulos G; "Fish oil fatty acid supplementation and active ulcerative colitis: A double-blind, placebo controlled crossover study," *The American Journal of Gastroenterology*, April 1992;87(4):432-433.
51. Shabert J; *The Ultimate Nutrient Glutamine*, Avery, Garden City Park, NY, 1994; Wilmore D; personal conversation Spring 1995.

52. Scheppach W, Sommer H, et al; "Effect of butyrate enemas on the colonic mucosa in distal ulcerative colitis," *Gastroenterology* 1992;103:51-56.
53. Scheppach W; "Effects of short chain fatty acids on gut morphology and function," *Gut* 1994(Suppl.1):s335-s38; Babbs CF; "Oxygen radicals in ulcerative colitis," *Free Radic Biol Med*, 1992;13(2):169-81; Gross V; et al; "Free radicals and inflammatory bowel diseases, pathophysiology and therapeutic implications," *Hepato-Gastroenterology*, 1994;41:320-327.

Chapter 13. Natural Therapies for the Diverse Consequences of Faulty Digestion

1. Hippocrates, "On Airs, Waters, and Places," *The Genuine Works of Hippocrates*, Francis Adams Translator, London; Leslie P. Adams, Jr., 1849, quoted in *The Coming Plague*, by Laurie Garrett, Farrar, Straus & Giroux, New York, 1994, p.234.
2. Abyad A; Boyer JT, "Arthritis and Aging," *Curr Opin Rheumatol*, 1992 Apr;4(2):153-9.
3. Mulbert AE; et al; "Identification of Nonsteroidal antiinflammatory drug-induced gastroduodenal injury in children with juvenile rheumatoid arthritis," *Journal of Pediatrics*, April 1993;645-6.
4. Randolph T; Moss RW; *An Alternative Approach to Allergies*, Harper Perennial, New York,1990, p. 162.
5. Ibid., p. 279.
6. Darlington IG; "Dietary therapy for arthritis" *Nutrition and Rheumatic Diseases/Rheumatic Disease Clinics of North America*, May 1991;17(2):273-285.
7. Seignalet J; "Diet, fasting and rheumatoid arthritis," *The Lancet*, Jan 4 1993; 339:68-69.
8. Machtey I; "Vitamin E and Arthritis/Vitamin E and Rheumatoid Arthritis," *Arthritis and Rheumatism*, Sept 1991;34(9):1205.
9. McCarty MF; "The neglect of glucosamine as a treatment for osteoarthritis—a personal perspective," *Med Hypotheses*, 1994 May;42(5):323-7.
10. Bingham R; et al; *J. Applied Nutrition*, 1975; 27:45; 1978;30:127.
11. Nielsen GL; et al; "The effects of dietary supplementation with N-3 polyunsaturated fatty acids in patients with rheuma-

toid arthritis," *European J of Clin Investigation* 1992;22:687-691; McCarthy G; Kenny D; "Dietary fish oil and rheumatic diseases," *Seminars in Arthritis and Rheumatism*, June 1992;21(6):368-375.

12. Srivastava KC Mustafa T; "Ginger (Zingiber officinale) in rheumatism and musculoskeletal disorders, *Medical Hypothesis*, 1992;39:342-348.
13. Flynn M; "The effect of folate and cobalamine on osteoarthritis and hands," *J Amer College of Nutrition*, 1994;13(4):351-356.
14. Trock DH; et al; "A double-blind trial of the clinical effects of pulsed electromagnetic fields in osteoarthritis," *The Journal of Rheumatology*, 1993;20(3):456-460.
15. Barrie S; "Dysbiosis & Immune disease, a retrospective study," tape from lecture, Great Smokies, Asheville, NC,1993.
16. Galland, L; telephone conversation, June 1995.
17. Henricksson AEK; "Small intestinal bacterial overgrowth in patients with rheumatoid arthritis," *Annals of Rheumatic Diseases*, 1993;52:503-510
18. Machtey I; "Vitamin E and Arthritis/Vitamin E and Rheumatoid Arthritis," *Arthritis and Rheumatism*, Sept 1991;34(9): 1205.
19. Leventhal LJ; et al; "Treatment of rheumatoid arthritis with gamma-linolenic acid," *Annals of Internal Medicine*, Nov 1, 1993;119(9):867-873.
20. Brzeski M; et al; "Evening primrose oil in patients with rheumatoid arthritis and side-effect of non-steroidal anti-inflammatory drugs," *Brit J of Rheumatology*, 1991;30:370-372.
21. Keough C; *Natural Relief for Arthritis*, Pocket books, New York,1983, p.161; Disilvestro RA; "Effects of copper supplementation on ceruloplasmin and copper-zinc superoxide dismutase in free-living rheumatoid arthritis patients," *Journal of Amer College of Nutrition*, 1992;11(23):177-180.
22. Germain BF; "Silicone breast implants and rheumatic disease," *Bulletin on the Rheumatic Diseases*, Oct 1992;41(6):1-4.
23. Troughton PR; Morgan AW; "Laboratory findings and pathology of psoriatic arthritis," *Baillieres Clin Rheumatol*, 1994 May;8(2):439-63.
24. Leirisalo-Repo M; "Enteropathic arthritis, Whipple's Disease, juvenile spondyloarthropathy, and uveitis," *Curr Opin Rheumatol*, 1994;Jul;6(4):385-90.

25. Barrie S; "Dysbiosis & immune disease, a retrospective study," tape from lecture, Great Smokies, Ashville, NC,1993; Houston L; "Dietary change in arthritis," *The Practitioner* June 1994;238:443-448; Lahesmaa R; et al; "Molecular mimicry: Any role in the pathogenesis of spondyloarthropathies?," *Immunol Res*, 1993;12(2):193-208.
26. McSherry J; "Chronic fatigue syndrome: A fresh look at an old problem," *Can Fam Physician*, 1993 Feb;39:336-40.
27. Bland J; *Advancement in Clinical Nutrition* HealthComm 1993-4, p.3, taken from: Komaroff AL; "Clinical presentations of Chronic Fatigue Syndrome," *Chronic Fatigue Syndrome*, Bock and Whelan editors, Wiley and Sons Ltd., p.102-31.
28. Rigden S; Bland J; "UltraClear Program and Cost Effectiveness," 1993, *Advancement in Clinical Nutrition*, HealthComm, Gig Harbor, WA, 1993-4, p. 12.
29. Shafran SD; et al; "Chronic fatigue syndrome," *The Amer J of Medicine*, June 1991;90:730-740.
30. Lapp CW; Cheney P; "Chronic fatigue syndrome: self-care manual, Feb 1991," *The CFIDS Chronicle Physicians' Forum*, March 1991;1(1):14-17.
31. Gray JB; Martinovic AM; "Ecoisanoids and essential fatty acid modulation in chronic disease and chronic fatigue syndrome," *Medical Hypothesis*, 1994 July;;43:31-42.
32. "A follow-up on malic acid: CFIDS Buyer's Club," *Health Watch* Spring 1993;3(1):1,3.
33. Bland J; *Applying New Essentials in Nutritional Medicine*, HealthComm, Gig Harbor, WA, 1995, p. 8
34. Sampson HA; "The immunopathogenic role of food hypersensitivity in atopic dermatitis," *Acta Derm Venereol Suppl*, (Stockh) 1992;176:34-37
35. Sazanova NE; et al; "Immunological aspects of food intolerance in children during first years of life," *Pediatriia*, 1992:(3):14-18.
36. Kalimo K; "Yeast Allergy in Adult Atopic Dermatitis," *Immunological and Pharmacological Aspects*, 1991;4:164-167.
37. Werbach M; *Healing with Nutrition*, 1993, Harper Collins, New York, p. 124
38. Horrobin DF; "Fatty acid metabolism in health and disease: The role of delta-6 desaturase," *Amer J of Clinical Nutrition*,

1993;57(Suppl):732S-7S; Oliwiecki S; et al; "Levels of essential and other fatty acids in plasma and red cell phospolipids from normal controls in patients with atopic eczema," *ACTA Derm. Venereol Stockholm*, 1990;71:224-228.

39. Soyland E, et al; "Dietary supplementation with very long-chain omega-3 fatty acids in patients with atopic dermatitis," *Br J of Dermatology*, 1994;130:757-764.
40. Goldenberg DL; "Fibromyalgia, chronic fatigue syndrome, and myofascial pain syndrome" *Curr Opin Rheumatol*, 1994 Mar; 6(2):223-33; Briggs NC; Levine PH; "A comparative review of systemic and neurological symptomatology of 12 outbreaks collectively described as chronic fatigue syndrome, epidemic neuromyasthenia, and myalgic encephalomyelitis," *Clin Infect Dis*, 1994 Jan;18 Suppl 1:S32-42; Moldofsky H; "Fibromyalgia, sleep disorder and chronic fatigue syndrome," *Ciba Found Symp*, 1993;173:262-71; discussion 272-9.
41. Bland J; *Applying New Essentials in Nutritional Medicine*, Health-Comm 1995, p. 14.
42. Galland L; telephone conversation, June 1995.
43. Moreshead J; Jaffe R; "Fibromyalgia: Clinical success through enhanced host defenses: A case-controlled outcome study," *IAACN Syllabus* Sept 1994 Dallas TX.
44. Bland, *Applying New Essentials*, pp. 15-19.
45. Eisinger J; et al: "Glycolysis abnormalities in fibromyalgia," *J of the Amer College of Nutr*, 1994;13(2):144-148.
46. Kimikazu I; Kiyonana J; Ishikawa M; "Studies on Gamma-Oryzanol II: The Anti-Ulcerogenic Action, Research Institute," Otsuka Pharmaceutical Co., Ltd, Tokushima 771-01; Resnick C; "The Effects of Gamma-Oryzanol on Ulcers, Gastritis, Hyperlipidemias, and Menopausal Disorders", *Research Review*, Tyler Encapsulations, Gresham, OR, 1993.
47. Yunus MB; et al; "Plasma tryptophan and other amino acids in primary fibromyalgia: a controlled study," *J of Rheumatology*, 1992;19:1:90-94.
48. McCarty DJ; et al; "Treatment of pain due to fibromyalgia with topical capsaicin: A pilot study," *Seminars in Arthritis and Rheumatism*, June 1994;(Suppl 3):23(6):41-47.
49. Grassetto M; Varotto A; "Primary fibromyalgia is responsive to S-adenosyl-L-methionine," *Current Therapeutic Research*, July

1994;55(7):797-806; Ianniello A; et al; "S-adenosyl-L-methionine in Sjogren's syndrome and fibromyalgia," *Current Therapeutic Research*, June 1994;55(6):699-706.

50. Deluze C; et al; "Electroacupuncture in fibromyalgia: Results of a controlled trial," *Br Medical Journal*, Nov 21, 1992; 305:1249-1251; Faivelson, S; "Electroacupuncture tried for pain of fibromyalgia," *Medical Tribune*, Dec 24, 1992;24.
51. Marks, David and Laura, "Food Allergy: Manifestations, Evaluation and Management," Food Allergy, *Postgraduate Medicine*, Feb. 1, 1993:93(2):191-201.
52. Golos N; Golos-Golbritz F; *If it's Tuesday it Must be Chicken*, Keats Publ., New Canaan, CT, 1979.
53. *Elisa/Act Patient Handbook*, Serammune Laboratories, Reston, VA.
54. Stewart WF: Schechter A; Rasmussen BK; "Migraine prevalence. A review of population-based studies," *Neurology*, 1994 June;44(6 Suppl 4):S17-23; Silberstein SD; Lipton RB; "Epidemiology of migraine," *Neuroepidemiology* 1993;12(3):179-84; Smith R; "Chronic headaches in family practice," *J Am Board Fam Pract* 1992 Nov-Dec;5(6):589-99.
55. Robbins L; "Precipitating factors in migraine; a retrospective review of 494 patients," *Headache* 1994 Apr;34(4):214-6.
56. Jones M; "Migraine Headaches and Food," *NOHA News*, Spring 1989 XIV, No. 2.
57. Martelletti P; "T cells expressing IL-2 receptor in migraine," *Acta Neurol* 1991 Oct;13(5):448- 56.
58. Murray M; Pizzorno J; *The Encyclopedia of Natural Medicine*, Prima, Rocklin, CA, 1991, p. 416.
59. Heuser G, et al: "Candida Albicans and migraine headaches: a possible link," *Journal of Advancement in Medicine*, Fall 1992; 5 (3): 177-187.
60. Soriani S; et al; "Serum and red blood cell magnesium levels in juvenile migraine patients," *Headache*, 1995 Jan;35(1):14-16; Thomas J; et al; "Migraine treatment by oral magnesium intake and correction of the irritation of buccofacial and cervical muscles as a side effect of mandibular imbalance," *Magnes Res*, 1994 Jun;7(2):123-7; Taubert K; "Magnesium in migraine. Results of a multicenter pilot study," *Fortschr Med*, 1994 Aug 30;112(24):238-30.

61. Johnson ES; Kadam NP, Hylands DM and PJ; "Efficacy of feverfew as prophylactic treatment of migraine," *British Medical Journal*, 1985 Aug 31;2291(6495):569-73.
62. Schoenen J, et al.; "High-dose riboflavin as a prophylactic treatment of migraine: results of an open pilot study," *Cephalagia*, 1994;14:328-329.
63. Hesse J; et al; "Acupuncture versus metoprolol in migraine prophylaxis: A randomized trial of trigger point inactivation," *Journal of Internal Medicine*, 1994;235:451-456; Hesse J; et al; "Acupuncture compared with metoprolol for migraine gave mixed results," *Annals of Internal Medicine*, May 1994;235:451-6.
64. Black M; "Nicotinic acid and headache," *Cortlandt Forum* Aug 1990:26-30.
65. Anthony M; "Platelet superoxide dismutase in migraine and tension type headaches," *Cephalalgia*, 1994;14:1818-3.
66. Murray M; Pizzorno J; *The Encyclopedia of Natural Medicine*, Prima, Rocklin, CA, 1991, p. 487-91.
67. Alteras I; "The incidence of skin manifestations by dermatophytes in patients with psoriasis," *Mycopathologia*, 1986 Jul;95(1):37-9.
68. Menzel I; Holzmann H; "Reflections on seborrheic scalp eczema and psoriasis capillitii in relation to intestinal mycoses," *Z Hautkr*, April 1986;61(7):451-4.
69. Gaston L; et al; "Psychological stress and psoriasis: Experimental and prospective correlation studies," *ACTA Derm. Venereol*, 1991;(Suppl 156):37-43.
70. Fahrer H; et al; "Diet and fatty acids: can fish substitute for fish oil?" *Clin Exp Rheumatology*, 1991 July-Aug;9(4):403-6; Grimminger F; et al; "A double blind randomized, placebo controlled trial of N-3 fatty acid based lipid infusion in acute, extended guttate psoriasis: Rapid improvement of clinical manifestations and changes in neutrophil leukotriene profile," *Clinical Investigator*, 1993;71:634-643; Corrocher R; et al; "Effect of fish oil supplementation on erythrocyte lipid pattern, malondialdehyde production and glutathione-peroxidase activity in psoriasis," *Clin Chim Acta*, 1989 Feb 15;179(2):121-31; Berbis P, et al; "Essential fatty acids and the skin," *Allergy Immunology*, 1990 June;22(6):225-31.

71. Dochao A; et al; "Therapeutic effects of vitamin D and vitamin A in psoriasis: a 20 year experiment," *Actas Dermosfiliogr*, 1975;66(3-4):121-30.
72. Sukenik S; et al; "Treatment of psoriatic arthritis at the Dead Sea," *J Rheumatology*, 1994 Jul;21(7):1305-9.
73. Michaelsson G; Ljunghall lK; "Patients with dermatitis herpetiformis, acne, psoriasis, and Darier's disease have low epidermal zinc concentrations," *Acta Derm Venereol*, 1990;70(4):304-8; Leung RS; et al; "Neutrophil zinc levels in psoriasis and seborrheoic dermatitis," *Br J Dermatol*, 1990 Sep; 123(3):319-23; McMillan EM; et al; "Diurnal stage of circadian rhythm of plasma zinc in healthy and psoriatic volunteers," *Prog Clin Biol Res*, 1987;227B:295-303.
74. Michaelsson G; Berne B; et al; "Selenium in whole blood and plasma is decreased in patients with moderate and severe psoriasis," *Acta Derm Venereol*, 1989;69(1):29-34.
75. Harvima RJ; et al; "Screening of effects of selenomethionine-enriched yeast supplementation on various immunological and chemical parameters of skin and blood in psoriatic patients," *Acta Derm Venereol*, 1993 Apr;73(2):88-91;Fairris GM Lloyd B; et al; "The effect of supplementation with selenium and vitamin E in psoriasis," *Ann Clin Biochem*, 1989 Jan;26(Pt 1):83-8.
76. Ellis C; "Hot pepper cure: Capsaicin relieves psoriatic itch," *Modern Medicine* 1993:61:31 and Ellis C; *Journal of Amer Academy of Dermatology* Sept 1993; Kurkcuoglu N; Alaybeyi F; "Topical capsaicin for psoriasis," *B J of Dermatology*, Oct 1990;123(4):549-50.
77. Murray M; Pizzorno J; *The Encyclopedia of Natural Medicine*, Prima, Rocklin, CA, 1991, p. 488.
78. Hendel L; et al; "Esophageal candidosis in progressive systemic sclerosis: Occurrence, significance, and treatment with fluconazole," *Scand J Gastroenterol*, 1988 Dec;23(10):1182-6.
79. Gabay C; Kahn MF: "Male-type scleroderma: the role of occupational exposure," *Schweiz Med Wocheschr*, 1992 Nov 14;122(46):1746-52; Czirjak L; et al; "Localized scleroderma after exposure to organic solvents," *Dermatology*, 1994;189(4):399-401; Pelmear PL; Roos JO: Maehle WM; "Occupationally-induced scleroderma," *J Occup Med* 1992 Jan;34(1):20-25.
80. Sanchez-Roman J; et al; "Multiple clinical and biological auto-

immune manifestations in 50 workers after occupational exposure to silica," *Ann Rheum Dis*, 1993 Jul;52(7):534-8.

81. Lamm SH; "Silicone breast implants and long-term health effects: When are data adequate?" *J Clin Epidemiol*, 1995 Apr;48(4):507-11.
82. Strong AMM; et al; "The effect of oral linoleic acid and gamma-linolenic acid(Efamol)," *British J Clin. Practice*, 1985 Nov/Dec;39(11-12):444-445.

Index

D

E

G